Tunable Ferroelectric Matching Networks
implemented into High Power RF Amplifiers for High
Dynamic and Wideband Efficiency

Tunable Ferroelectric Matching Networks implemented into High Power RF Amplifiers for High Dynamic and Wideband Efficiency

Vom Fachbereich Elektrotechnik und Informationstechnik
der Technischen Universität Darmstadt
zur Erlangung der Würde eines
Doktor-Ingenieurs (Dr.-Ing.)
genehmigte

Dissertation

von

Dipl.-Phys.
Alex Wiens
geboren am 20.11.1984
in Frunse, Kirgisistan

Referent:	Prof. Dr.-Ing. Rolf Jakoby
Korreferent:	Prof. Dr.-Ing. Wolfgang Heinrich

Tag der Einreichung:	31.10.2016
Tag der mündlichen Prüfung:	22.02.2017

D17
Darmstadt 2017

Bibliografische Information der Deutschen Nationalbibliothek
Die Deutsche Nationalbibliothek verzeichnet diese Publikation in der
Deutschen Nationalbibliografie; detaillierte bibliografische Daten sind im Internet
über http://dnb.d-nb.de abrufbar.
1. Aufl. - Göttingen: Cuvillier, 2017
   Zugl.: (TU) Darmstadt, Univ., Diss., 2017

   ISBN 978-3-7369-9511-6
   eISBN 978-3-7369-8511-7

Preface

...to my father, who motivated me to pursue my doctorate...

This thesis presents the research results performed as a Ph.D. candidate in the institute for microwave engineering and photonics at Technische Universität Darmstadt. The research performed within the scope of this thesis was, doubtlessly, not possible without many of my colleagues, cooperation partners and friends whom I would like to acknowledge.

My sincere thanks go to my supervisor Prof. Dr.-Ing. Rolf Jakoby, who motivated and guided me from the very first day; the trust and patience he put into me have allowed me to grow, not only in a scientific but also in a personal way. I would also like to thank Prof. Dr.-Ing. Wolfgang Heinrich. His kind advices and suggestions have allowed me to explore the topics in my dissertation beyond my horizons. Prof. Dr.-Ing. Christian Damm is highly acknowledged for the time and work he spent with me at the institute.

My special thanks go to:

- Dr.rer.nat. Joachim R. Binder, Dr.-Ing. Christian Kohler and Dr.-Ing. Andreas Friederich from Karlsruhe Institute of Technology for the trustworthy cooperation and for providing the tunable material and samples.

- My dear friend, colleague and cooperation partner Dr.-Ing. Olof Bengtsson and Sebastian Preis from Ferdinand Braun Institute. Without both of you my work would not have been as successful as it became. Your support and advices were invaluable to me.

- My dear friend, former colleague and mentor Dipl.-Inf. Frank Lischke, who taught me to control first impulses and focus on important things to achieve my goals.

I am indebted by so many gifted people, who helped me during my work, starting with my colleague and friend Dr.-Ing. Martin Schüßler. Former colleagues and friends Dr.-Ing. Matthias Hansli, Dr.-Ing. Alexandar Angelovski, Dr.-Ing. Matthias Höfle and Saygin Bildik. My dearest friend, fellow student and institutes' sunshine cand. Dr.-Ing. Matthias Jost, who constantly managed to cheer me up on cloudy days – thank you!

My thanks also go to my Master students, especially Christian Schuster and Felix Lenze, I learned a lot supervising them. Also, I want to thank my new colleagues, who recently started for the fruitful off-topic discussions – Roland Reese, Sönke Schmidt and Matthias Nickel.

The technical staff and their expertise is greatly acknowledged in this work. Peter Kießlich, Andreas Semrad and Karin Boye: thank you so much for the work you do and have done during my time at the institute! Without you, a great part of this work would not have been possible.

Last but not least, I also want to thank Dr.-Ing. Holger Maune, who helped me with his knowledge, and thus, allowed me to overcome my limits.

I also need to thank my family and relatives. Doubtlessly, the outcome of my work would not have been the same without their support and motivation on my stony and long road towards the Ph.D.

Darmstadt, 2017

Alex Wiens

Kurzfassung

In dieser interdisziplinären Arbeit wurden die dielektrischen Eigenschaften von kontinuierlich steuerbaren ferroelektrischen Materialien hinsichtlich ihrer Eigenschaften im Hochleistungsbetrieb untersucht. Die dabei durchgeführten Untersuchungen reichen von der Materialebene, über die Modellierung der Komponenten bis hin zur komplexen Synthese von steuerbaren Hochfrequenzschaltungen.

Im Rahmen dieser Arbeit wurde ein neuartiger CAD-Ansatz zur Modellierung steuerbarer Kompositkeramiken entwickelt. Der Fokus wurde auf die Reduktion der dielektrischen Verluste gelegt, welche durch die Vermischung eines steuerbaren, ferroelektrischen Materials mit einem nicht steuerbaren Dielektrikum höherer Güte erreicht werden sollte. In dem vorgeschlagenen Modell wurde ein zusätzlicher Freiheitsgrad (*Kornüberlapp*) eingeführt, woduch eine, im Vergleich zu bereits etablierten Methoden, akkuratere Beschreibung von zweiphasigen Kompositmaterialien ermöglicht wurde. Der Vergleich zwischen Simulation und Messung zeigte gute Übereinstimmung, welche durch keine der zuvor verwendeten Modelle erreicht wurde.

Erste systematische Untersuchungen von steuerbaren, auf Ferroelektrika basierten, Parallelplattenkondensatoren (*engl.* MIM) und uniplanaren Interdigitalkondensatoren (*engl.* IDCs) für Hochleistungsanwendungen wurden durchgeführt. Dabei wurde das hohe Potential dieser Technologie offengelegt. Eine Leistungsverträglichkeit von mehr als $50\,\mathrm{W}/\mathrm{mm}^2$ wurden beobachtet. Die Linearität lag bei über $60\,\mathrm{dBm}$ (OIP3). Fortgeschrittene numerische Modelle von Komponenten wurden implementiert und ermöglichen eine direkte Implementierung in Schaltungssimulatoren zur Synthese steuerbaren Gesamtschaltungen.

Zu guter Letzt wurden die untersuchten Komponenten als Teil eines kompakten, steuerbaren Anpassnetzwerks in einem Transistorgehäuse untergebracht. Die entwickelten Modelle erlaubten eine drastische Reduktion der Einfügeverluste des synthetisierten Anpassnetzwerks unterhalb $1\,\mathrm{dB}$. Eine hybride Implementierung mit GaN HEM (*engl.* gallium nitride high electron mobility) Transistorzellen wies den bisher höchsten Wirkungsgrad ($\geq 60\,\%$ bei $2\,\mathrm{GHz}$) unter Verwendung von steuerbaren, ferroelektrischen Komponenten auf.

Abstract

In this interdisciplinary work, the dielectric properties of continuously tunable ferroelectric materials have been investigated with regard to their high frequency and high power handling capability. The investigations range from material level over component modeling to sophisticated tunable network synthesis methods.

Within this work, a novel CAD approach for the modeling of tunable composite ceramics has been developed with the goal to reduce the dielectric losses of tunable materials by combining them with high-Q non-tunable dielectrics. The model introduces an additional degree of freedom (*grain interconnection*) to accurately map the behavior of two phase composite ceramics. A comparison between simulation and measurements show good agreements, which previous models are unable to achieve.

First systematic investigations of tunable ferroelectric varactors in parallel-plate configuration (MIM) and inter-digital-capacitors (IDCs) for high power applications were performed, showing high potential of the components for the implementation in tunable high RF power circuits. A power handling capability of $\geq 50\,\mathrm{W/mm}$ has been observed with a linearity of OIP3 $\geq 60\,\mathrm{dBm}$. To allow the implementation of tunable components based on ferroelectric material, advanced numerical models have been implemented, allowing accurate prediction of the component's properties. The tunable component models enable easy deployment in a circuit simulator and allow accurate prediction of tunable circuit behavior.

The elements were implemented as part of a compact tunable impedance matching network inside a transistor package. The accurate models allowed a tremendous reduction of insertion losses of the synthesized matching networks down to below $1\,\mathrm{dB}$. Hybrid implementation with gallium nitride high electron mobility transistor cells showed the so far highest power added efficiencies ($\geq 60\,\%$ at $2\,\mathrm{GHz}$) achieved with tunable ferroelectric circuits.

Contents

1 Introduction   11

2 Fundamentals of Nonlinear Dielectrics   15
    2.1   Electric Polarization 15
    2.2   Perovskite Ferroelectrics 17
    2.3   Barium Strontium Titanate 20
       2.3.1   Modeling of Electric Tunability 24
       2.3.2   Process Technology for Tunable Components 26

3 Ferroelectric Composite Material and Varactor Optimization   31
    3.1   Multi Dielectric Composites 31
       3.1.1   Modeling of Dielectric Properties 32
       3.1.2   Processing and Characterization 47
       3.1.3   Comparison of Models and Measurement Results 51
    3.2   Varactor Topologies: Design and Optimization 56
       3.2.1   Uniplanar Thick Film Varactors 57
       3.2.2   Multilayer Thick- and Thin Film Varactors 65
       3.2.3   External Biasing Concepts: Resistive and Inductive Decoupling   67

4 Large Signal Characteristics of Ferroelectric Varactors   75
    4.1   Thermal Conductivity of Barium Strontium Titanate 76
    4.2   RF-induced Self Heating 78
       4.2.1   Modelling of Thermal Behaviour 79
       4.2.2   Large Signal Measurements 82
    4.3   Non-linear Characteristics 89
       4.3.1   Models 89
       4.3.2   Intermodulation Measurements and Comparison 93

5 Integration of Tunable Matching Networks into RF Power Amplifiers   95
  5.1 Transistor Characterization for Impedance Matching 99
  5.2 Filter-based Matching Network Synthesis 102
      5.2.1 Thick Film Implementation 112
      5.2.2 Thin Film Implementation 116
  5.3 CAD-aided Impedance Space Matching Network Synthesis 118
  5.4 Hybrid Implementation into a GaN HEMT 125

6 Conclusion and Outlook   135

A Mathematical Appendix   139

B Characterization Appendix   143

C Figure Appendix   151

D Table Appendix   153

E Lithography Appendix   155

F Measurement equipment   157

Symbols and Abbreviations   157

Bibliography   163

Curriculum Vitae   183

1 Introduction

The continuous progress in modern wireless communication systems during the last decades is, and will be driven by the steadily increasing number of wireless devices and services. Beside the usual consumer electronics with the growing demand for broadband access to the internet, other upcoming fields such as *Industry 4.0* [Sch+15], *Internet of Things* (IoT) [Wor] as well as massively connected sensor networks [Com] are crowding the available frequency spectrum. Over the years, new bands have been allocated and advanced modulation schemes have been developed to cope with this circumstance. While adjusting the front end to the new requirements within the digital domain is rather straight forward, the analogue part of a transceiver can be considered as rigid, and therefore, the limiting factor of the whole wireless system. For example, the utilized filters and matching networks have an inherent trade-off between the achievable transmission and the bandwidth [Bod55], and are therefore, optimized to perform best in a certain, predefined frequency range. Especially in mobile devices, antennas play a significant role and are the focus of intense optimization to allow broadband operation (usually at the expense of matching). The amplifiers on the other hand are also limited in their application field, due to the inherent non-trivial matching problem. Deviation from the pre-defined operation scenarios (frequency, input power, temperature, etc.), usually leads to a drastic drop in performance and efficiency of the devices. Consequently, one of the key elements for multi band and multi standard operation is the possibility to adjust the analogue part of the RF front end by tuning the RF components.

To overcome the intrinsic physical limitations, which are partially addressed within this work, the implementation of modern transceivers is usually carried out in parallel trunks. Each branch is optimized to perform best at distinct standards within a certain, specified frequency band. However, this implementation suffers from the fact that only predefined standards and band as well as preequipped trunks can be used, which eventually leads to overaged transceivers, that need to be replaced. Further, the parallel implementation not only requires larger space and volume, but is also accompanied with an increased amount of components, which directly affects the costs. In case of active components, parallel implementation affects the

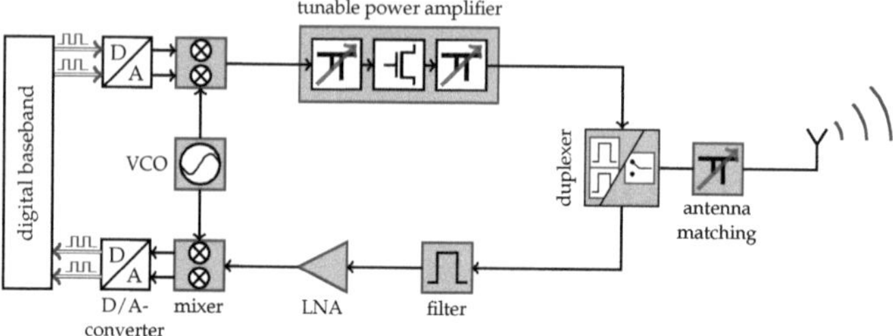

Figure 1.1: Concept of a fully reconfigurable RF front end in accordance to [Sch07a]. Gray-shaded blocks denote tunable elements such as matching networks, filters and other components.

overall power consumption, and hence, the efficiency. In the last consequence, this architecture requires isolated bands, which leave unused frequencies within the transitions from one band to another.

Especially the realization of *software defined radios* (SDR) and *cognitive radios* requires a completely reconfigurable analogue RF front end, which is not bound by predefined frequency bands and standards [Mit95; Mit00; PLL08]. First applications and devices, where exemplary the antenna's impedance is adjusted by tunable circuits have attracted attention [Gon+14]. Tunable filters, phase shifters and other components have been investigated and realized in various technologies for various operation frequencies [Gae+09; Zhe11; CC16]. However, the fully reconfigurable transceiver as depicted in Fig. 1.1 has not yet been deployed due to heterogeneous technologies, frequencies etc. Another important fact is that the energy efficiency, especially in handheld devices is one of the most critical requirements. In this context, the power amplifier (PA) of a transceiver system, unambiguously, can be considered as the main power consumer with around 70 % of the total transceiver's consumption [BZB10]. Depending on the operation class of the PA, large fraction of this energy is basically converted into heat (e.g. in class-A operation at least 50 %). The remaining portion is ideally the amplified signal, where the task is either to match over a wide frequency range and accept a reduced efficiency as a consequence of poor matching, or to reduce the bandwidth and tune the matching to achieve high efficiency. Consequently, the design and development of a tunable, efficient power amplifier is of utmost importance for energy efficient software defined radios.

The tunable elements of building blocks depicted in Fig. 1.1 can be fabricated implementing semiconductor varactor diodes, micro electromechanical systems (MEMS), liquid crystals or ferroelectric tunable capacitors. Varactor diodes suffer from intrinsically low linearity and limited power handling capability for GHz RF applications, and hence have to be implemented in large arrays [Raa11] or can be only used in the receive path. MEMS offer good linearity and moderate power handling capability but suffer from mechanical fatigue [DSB08]. Liquid crystal based components have shown superior performance at frequencies above 10 GHz[Gae+09; Wei+13]. Ferroelectric ceramics such as KTN [Lau+05], BZT [XMI05] or BZN [Par+05] and many others have shown better performance below 10 GHz. Among them, especially ferroelectric varactors based on barium strontium titanate (BST) were found to have the most promising properties for the realization of tunable components at microwave frequencies [Tag+03; Gie09; Saz13].

The BST thick film technology offers a variety of benefits compared to other previously mentioned technologies. The material and the fabricated components offer sufficient and continuous tunability in combination with acceptable insertion loss below 10 GHz as well as good linearity. The material is inexpensive and allows the realization of all required building blocks for an SDR, especially in the high power path. Further, the bias voltage in combination with negligible leakage current allows powerless tuning, which has a direct impact on the efficiency of the system.

Within the last decade, individual tunable components for a fully reconfigurable RF front end based on BST have been demonstrated, including phase shifters [Gie+06; Ven07; Nik+14], filters [Nat+05; Sch16] and tunable impedance matching networks for antennas [Sch07b; Zhe11]. However, usually these components have only been tested under small signal conditions. First high power investigations have been conducted [Sch07a; Mau11] for inter-digital-capacitors (IDCs) and first tunable impedance matching networks have been investigated for high power transistors [Mau+12]. Nevertheless, the obtained efficiency and power handling capability of the fabricated components ($\geq$ 6 dB insertion loss) did not yield the expected improvements, which are necessary to advance the idea of a fully reconfigurable analogue RF front end.

The focus of this work is the systematic investigation of BST varactors for tunable impedance matching networks, compact enough to be implemented into the RF housing of a transistor. Especially the recently developed MIM varactors are examined for their suitability in high power applications.

Sophisticated numerical models for composite materials have been developed to allow CAD-aided prediction of dielectric properties and enable the design of novel, tailored materials.

Advanced numerical models of tunable components have been implemented and allow accurate prediction on tunability on a physical level and enable the extraction of key properties such as capacitance, resonance frequency as well as power handling capability. The designed models can be easily implemented in a circuit simulator and facilitate accurate prediction of tunable circuits behavior.

The proposed models are tested by two inherently different network synthesis methods, including filter theory design, and routes to consider tunable elements in the conventional (non-tuanble) filter synthesis. Further, a CAD-aided network synthesis method is proposed, which allows the synthesis of tunable circuits with exactly predefined characteristics.

2 Fundamentals of Nonlinear Dielectrics

2.1 Electric Polarization

A dielectric material is an electrical insulator that can be polarized by an externally applied electric field. When placed in an electric field, the charges, bound within the material, do not flow from one electrode to the other as they do in a conductor, but only slightly shift from their average equilibrium positions causing dielectric polarization $\vec{p}$, which is a measure of the separation of positive and negative electrical charges:

$$\vec{p} = q\vec{r} \ , \tag{2.1}$$

with q being the charge and $\vec{r}$ being the distance of the charges. Considering a macroscopic volume, the effective (macroscopic) polarization can be obtained by summing up the individual microscopic polarizations $\vec{p}$:

$$\vec{P} = \sum_i \vec{p}_i / V \ , \tag{2.2}$$

with V being the volume of the dielectric material. The polarization mechanisms can be subdivided depending on their origin:

- Space-charge polarization, where polarization occurs due to the diffusion of ions, along the field direction, thereby giving rise to redistribution of charges in the dielectrics.

- Orientation polarization, where molecules with built in dipole moments can independently rotate (and/or move) in an external electrical field.

- Ionic polarization, where ions in a lattice can be shifted from their equilibrium position by an external electrical field.

15

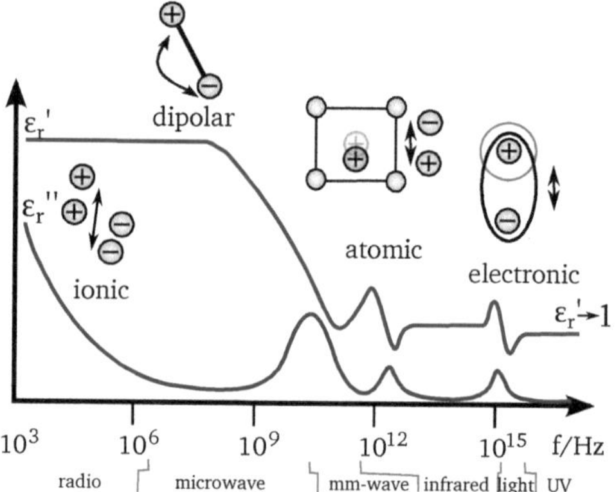

Figure 2.1: Qualitative frequency dependent real and imaginary part of the complex permittivity of a dielectric material, see (2.10).

- Electronic polarization, where the electrons, circling around an ion, are displaced from their average position with respect to the ion and, hence, create a dipole moment.

Depending on the frequency f of the external field $\vec{E}(f)$, the dominating mechanism is changing, due to different mass and mobility of the charges, as depicted in Fig. 2.1. In the simplest case[1], this phenomenon can be described as a mechanical spring-mass system, where the displacement $\vec{r}(t)$ of a charge can be written as:

$$\frac{q}{m_q}\vec{E}(t) - \frac{\delta^2}{\delta t^2}\vec{r}(t) - \Gamma\frac{\delta}{\delta t}\vec{r}(t) - \omega_0^2\vec{r}(t) = 0 \ , \tag{2.3}$$

with m_q being the mass of the charge, ω_0 the resonance frequency of the movement and Γ the damping of the movement. In the case of a harmonic excitation (2.3) can be written as:

$$\frac{q}{m_q}\vec{E}(\omega) + \omega^2\vec{r}(\omega) - j\omega\Gamma\vec{r}(\omega) - \omega_0^2\vec{r}(\omega) = 0 \ , \tag{2.4}$$

with the solution of the differential equation:

$$\vec{r}(\omega) = \frac{q}{m_q} \cdot \frac{\vec{E}(\omega)}{\omega_0^2 - \omega^2 + j\omega\Gamma} \ . \tag{2.5}$$

[1]More advanced models have been developed by Cole-Cole [CC41] and Debye [Cof04]

And with (2.1) the polarization can be expressed as:

$$\vec{p}(\omega) = \frac{q^2}{m_q} \cdot \frac{\vec{E}(\omega)}{\omega_0^2 - \omega^2 + j\omega\Gamma} \quad . \tag{2.6}$$

Using the relation between the displacement field $\vec{D}$ and the electric field:

$$\vec{D} = \underline{\varepsilon}\vec{E} = \varepsilon_0\vec{E} + \vec{p} \ , \tag{2.7}$$

a complex permittivity $\underline{\varepsilon}$ can be defined:

$$\underline{\varepsilon} = \varepsilon_0 \left(1 + \frac{q^2}{\varepsilon_0 m_q} \cdot \frac{1}{\omega_0^2 - \omega^2 + j\omega\Gamma} \right) = \varepsilon' - j\varepsilon'' \ , \tag{2.8}$$

where the imaginary part can be seen as dissipated energy and the real part can be considered as stored energy within the dielectric in the form of electrical field. Consequently, the ratio of both, can be used as a measure to characterize the quality of the material.

$$\tan\delta = \frac{\varepsilon''}{\varepsilon'} = \frac{1}{Q} \ , \tag{2.9}$$

where Q is referred to a quality factor of a dielectric.

$$\underline{\varepsilon} = \varepsilon' - j\varepsilon'' = \varepsilon' \left(1 - j\tan\delta \right) = \varepsilon_0\varepsilon_r \left(1 - j\tan\delta \right) \quad . \tag{2.10}$$

However, in absence of an external electrical field $\vec{E}$, most dielectric materials have randomly oriented dipole moments summing up to $\vec{P} = 0$. Ferroelectrics, on the other hand, belong to a sub-class of dielectrics, which allow the preservation of an internal polarization. The properties of this material class is the focus of the following section.

2.2 Perovskite Ferroelectrics

Ferroelectrics obtained their name due to their phenomenological accordance to ferromagnetics, which maintain their magnetic moments, once oriented in an external magnetic field. In a similar manner, the ferroelectric materials maintain a reminiscent macroscopic polarization P_R in the absence of an external electrical field. This effect is also often referred to as a *hysteresis* in polarization. Fig. 2.2 shows a qualitative polarization of a ferroelectric material versus external electrical field. In the

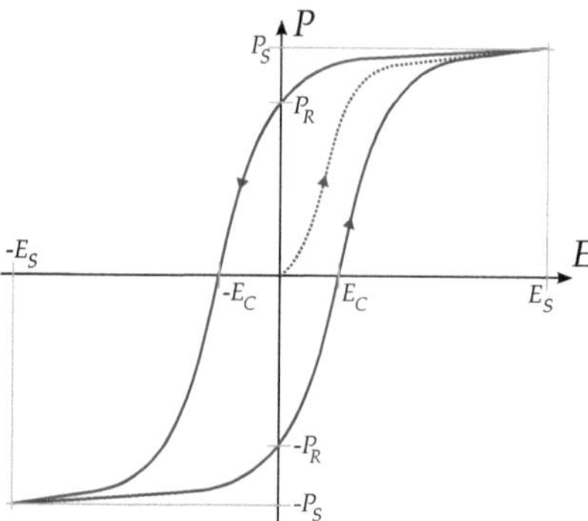

Figure 2.2: Qualitative characteristics of a ferroelectric polarization versus the external field.

initial state (red curve), the dipole moments $\vec{p}$ of the material are randomly oriented summing up to $\vec{P} = 0$. Application of an external field aligns the dipole moments until a saturation polarization P_S at a saturation field strength E_S is reached. After reducing the external field to $E = 0$, a certain reminiscent polarization P_R remains. The field has to be reversed to a value E_C to obtain $\vec{P} = 0$. Numerous ferroelectrics are found in a so called Perovskite structure, with the general formula of ABO_3. In this structure, an A-site ion, on the corners of the lattice, is usually an alkaline earth or rare earth element. The center of the lattice is occupied by B-site ions and can be 3d, 4d, and 5d transition metal elements. This material class distinguishes itself by a characteristic transition between a ferroelectric phase in a tetrahedral- and a paraelectric phase in a cubic lattice structure [Pau06]. Depending on the materials, this transition occurs at a distinct temperature called Curie temperature T_C. At this point, the permittivity ε' reaches a maximum. Fig. 2.3 qualitatively shows this transition, which can be thermodynamically described by the Ginzburg-Landau-Theory [GS01].

The nonlinear response of P manifests the E-field dependency of the relative permittivity:

$$\varepsilon_r = \frac{1}{\varepsilon_0}\frac{\partial P}{\partial E} \quad . \tag{2.11}$$

For the realization of tunable components, a paraelectric phase is preferred, as in this phase a memory effect of the polarization is avoided (or reduced). Usually

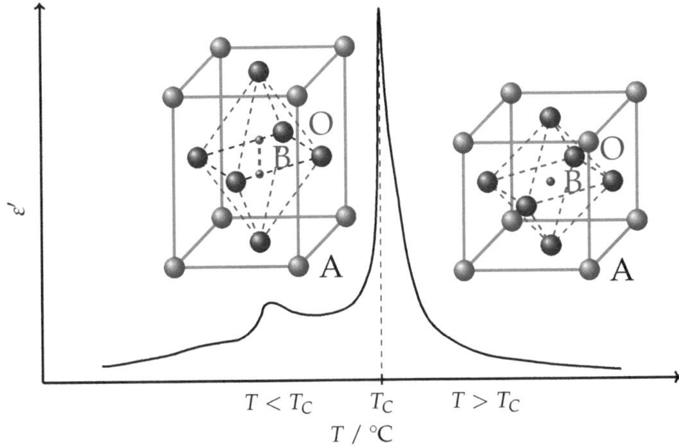

Figure 2.3: Qualitative characteristic of a Perovskite ferroelectrics versus temperature with a distinct transition from ferroelectric to paraelectric phase at the Curie temperature T_C. The inset shows a tetrahedral and cubic lattice structure without external bias fields. The axes are randomly scaled.

the material is operated closely above the Curie temperature T_C to maintain large electric tunability:

$$\tau(E) = \frac{\varepsilon_r(E = 0) - \varepsilon_r(E)}{\varepsilon_r(E = 0)} \quad , \tag{2.12}$$

where E is the electric field. For tunable varactors (2.12) is less convenient and rewritten as capacitances:

$$\tau(U_B) = \frac{C(U_B = 0) - C(U_B)}{C(U_B = 0)} \quad , \tag{2.13}$$

where U_B is the applied bias voltage. However, at the Curie temperature, not only the tunability τ is highest but also the dielectric losses of the material $\tan \delta$. Therefore, the choice of the material system and the operation temperature are of importance. There are also applications below Curie temperature, where the memory effect of the polarization is used to store information, such as *FeRAM* [Zhe11; Sco07].

However, for the work here, the paraelectric phase of the ferroelectric materials is of predominant interest.

One of the most prominent ferroelectric materials for microwave applications is barium strontium titanate (BST) with barium or strontium as the at the A-site ions and titanium as the B-site ion [Gev09]. In the following, the focus is put on its fabrication, processing, dielectric properties and modeling.

2.3 Barium Strontium Titanate

Commercial devices are usually operated in a temperature range between $0\,°C$ and $50\,°C$. Hence, the Curie temperature of the tunable material should be below or around the lower temperature limit to keep the paraelectric phase throughout the desired temperature range. So far, no pure ferroelectric material with Perovskite structure is known that has a Curie temperature of approximately $0\,°C$. Therefore, polycrystalline materials are synthesized, where the phase transition from ferro-electric to paraelectric is adjusted by the amount of the constituents. Pure barium titanate ($BaTiO_3$) has a transition temperature at around $T_{C,BaTiO_3} \approx 105\,°C$, which can be reduced by an admixture of strontium titanate ($SrTiO_3$) with a transition temperature of $T_{C,SrTiO_3} \approx -250\,°C$ [Tag+03]. In this mixture, the Sr ions substitute the Ba ions at the A-site. The result is a polycrystalline Perovskite ferroelectric $Ba_xSr_{1-x}TiO_3$, where the Curie temperature can be adjusted by the stoichiometric factor $x \in \{0..1\}$. Other rare earths can be combined in a similar manner to adjust the transition temperature but with inferior properties at microwave frequency range [Lau+05; XMI05; Par+05; Pau06]. Fig. 2.4 shows the permittivity ε_r versus temperature for different stoichiometric compositions x of bulk ceramic materials, measured at $1\,kHz$ [Jeo04]. With increasing Ba fraction T_C is shifted towards higher temperatures. At the same time the permittivity is decreasing. For single crystal and bulk ceramic materials, the transition temperature T_C can be approximated as [Tag+03]:

$$
\begin{aligned}
T_{C,s}(x) &= -244.8 + 368.3x - 18.58x^2\,[°C]\,, &\quad \text{Single crystal} \\
T_{C,c}(x) &= -240.9 + 485.6x - 136.8x^2\,[°C]\,, &\quad \text{Ceramic}
\end{aligned}
\tag{2.14}
$$

The single crystal shows a nearly linear dependence between x and the resulting Curie temperature, while bulk ceramics show increased cubic dependence. The effect can be attributed to defects within the material. According to (2.14), a Curie temperature around $0\,°C$ is obtained with $x = 0.6$, which is in good agreement with reported measurement results [Jeo04; Tag+03]. Consequently, the material is

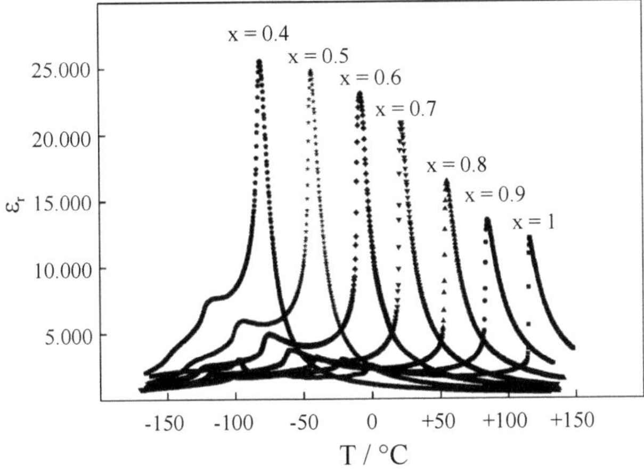

Figure 2.4: Temperature dependence of ε_r of a bulk ceramic for different stoichiometric compositions x measured at 1 kHz [Jeo04].

in paraelectric lattice structure throughout the required temperature. This holds especially at room temperature $T \approx 20\,°C$, where the material and components will be characterized. However, the relatively high permittivity of bulk ceramics ($\varepsilon_r \gg 1000$ at room temperature) hinders the fabrication, and also in addition requires strong miniaturization of the components, which makes this material form impractical for the targeted frequency range of several GHz. To target this circumstance, the permittivity can be reduced by different processing techniques. Fig. 2.5 briefly summarizes the processing steps for bulk ceramics, thick and thin films. In addition, the sol-gel route [Pau06] is presented, which was used for the fabrication of $Ba_{0.6}Sr_{0.4}TiO_3$ in this work. One advantage of this method is the simple possibility to add doping elements, which are used to modify the dielectric properties of the BST [Pau06; Zho12]. Further, this synthesis method allows flexible volume scaling. More details on the fabrication of precursors and ceramic powders can be found in [Pau06; Zho12; Koh16; Gev09]. Depending on the targeted application, the calcinated powder can be further ball milled in acetone with zirconia grinding media and mixed with other ceramics [Cha+08].

The obtained ceramic powder can be formed by pressing dies into pellets and sintered at temperatures above 1200 °C. Usually, the sample shrinks during that

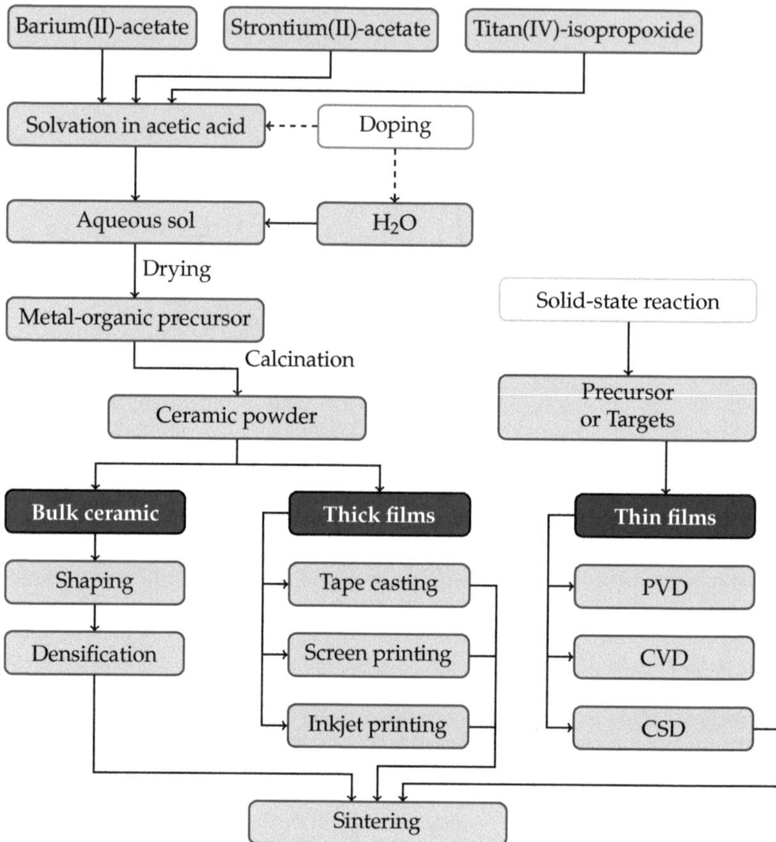

Figure 2.5: Processing routes of bulk and thick film BST ceramics in accordance to [Pau06] with different processing techniques. Brief summary of the thin film route and processing in accordance to [Gev09].

process by up to 30 % in volume with a theoretical porosity below 1 %. The resulting sample can be afterwards diced, milled or lapped to meet the specific geometric requirements. In this work, the fabricated bulk samples were lapped to 0.5 mm thick disks and processed with silver electrodes, forming a parallel-plate capacitor, see Chapter 4.

For thick film applications, the grinded powder (17 %) is diluted in terpineol (77 %) and Hypermer HD1 (1 %) and homogeneously dissolved. The viscosity of the resulting paste was adjusted by ethyl cellulose (5 %). The resulting paste is screen printed through a wire mesh with an approximate layer thickness of 4 µm and subsequently sintered at 1200 °C. The porosity of the sintered layer is in the range of 25 % − 35 %. The grain size and the porosity, both, influence the permittivity in a way that it significantly reduces ε_r, compared to bulk ceramics [Koh16]. After the sintering, metal electrodes are processed on top of the BST layer.

The pastes for ink-jet printing are fabricated and processed in a similar way. Two major differences are: the ceramic powder is more extensively milled to obtain smaller particles and the solid fraction in the ink is reduced to 5 %. This results in a smaller grain size distribution and higher porosity and further reduces the permittivity, compared to screen printed layers [Fri14]. Further, a minimum layer thickness of approximately 1.2 µm can be achieved with this process.

Two main doping routes have used within this work. The iron-fluorine (Fe-F) co-doped BST, which is found to have the highest Q [Zho12] and copper-fluorine (Cu-F) co-doped BST which has the effect of increased permittivity and increased sintering activity, which leads to reduced sintering temperatures. The latter is required for multilayer components. Typically, the thick films and bulk ceramics are sintered at $T \approx 1200$ °C, which exceeds the melting point of typical metals used for microwave devices, and hence allow only patterning of planar structures after the sintering step. This material doping, in combination with $ZnO\text{-}B_2O_3$ sintering additives, allows the reduction of the sintering temperature down to 850 °C, which allows the use of common electrode metals (gold, silver, etc), during the sintering step. This material, for the first time, allowed the fabrication of BST multilayer components and structures [Koh16; Fri14]. Multilayer BST components are investigated with regard to their high power properties in this work.

The thin films are usually processed by sputtering or pulsed laser deposition (PLD) and result in very high quality films [Gev09]. Consequently, they offer better Q and tunability $\tau(E)$. The layer heights are typically in the range of several 100 nm and, hence, require less bias voltage, compared to thick films or bulk. This technology has been commercialized by STMicroelectronics [STM15]. The fabrication of thin layers is usually limited to several hundred nanometer due to lattice mismatch between the bottom substrate (electrode) and the BST film, which leads to defects. However, recent work has shown that the lattice mismatch can be reduced by the use of oxide electrodes and thinner BST layers are feasible [Rad+14]. Films fabricated by chemical deposition methods usually require heat treatment in the range of 800 °C [Gev09].

Despite the fact that thin films are close to single crystalline films and have high material figure of merit (FoM) η:

$$\eta(E) = \frac{\tau(E)}{\tan \delta_{\max}} \, , \tag{2.15}$$

its applicability at RF frequency is usually limited by acoustic resonances. These resonances are the consequence of the piezoelectric effect, in combination with high quality lattice structure and the geometric dimensions [Gev09]. In thick films, so far, no obvious acoustic resonance has been observed due the polycrystalline, randomly oriented grains, which themselves hinder the propagation of acoustic waves. Fig. 2.6 summarizes the permittivity ε_r of the different technologies versus the applied bias voltage U_B.

Due to the large dimensions of the bulk ceramic materials, high bias voltage is required ($U_B \geq 1000\,\text{V}$) to tune the varactors. This material type also offers the highest permittivity and hence high electric tunability can be expected. The porous thick film has a reduced permittivity but requires significantly smaller bias voltage (but higher tuning fields) to achieve sufficient tunability, here 400 V. For the case of the commercial thin film varactors, the permittivity as well as the bias field have been roughly estimated, assuming a BST layer thickness of approximately 200 nm. The exact thickness is not known, therefore, the bias field as well as the absolute permittivity can vary up to a factor of 2. Though the material itself comes closest to single crystal, it is assumed that the relatively low permittivity in the range of 200 to 500 is caused by internal strain [Far+06]. Due to different characterization methods, which also involve different frequencies, the plot of $\tan \delta$ is omitted.

2.3.1 Modeling of Electric Tunability

Based on the theory of Ginzburg and Landau [GS01], several models have been proposed to mathematically describe the electric and thermal tunability of the nonlinear dielectrics. Vendig [VTZ98] extended the Ginzburg-Landau model to include the crystal quality, and accurately describes the nature of nonlinear dielectrics above Curie temperature [Mau11]. Chase [CCY05] simplified the model to describe the electric tunability, especially for thin films. Weil [Wei03] adapted the approximation

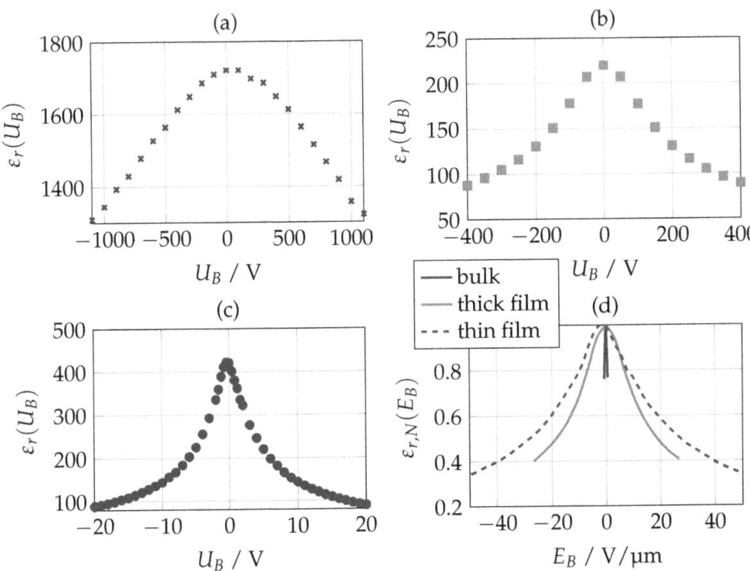

Figure 2.6: Measured permittivity ε_r versus the applied bias voltage U_B for different BST technologies - (a) Fe-F codoped BST bulk ceramic sintered at 1250 °C (13 MHz), (b) Fe-F codoped BST thick film sintered at 1200 °C (5 GHz), (c) thin film varactor from STMicroelectronics (2 GHz). Graph (d) shows the comparison with normalized permittivity $\varepsilon_{r,N}$ versus bias field E_B for each technology.

[Bet69] of the tunability, to describe the bias voltage dependent permittivity $\varepsilon_r(E)$ of ferroelectric thick films in the paraelectric phase:

$$\tau_\varepsilon(E) = \tau_{E_A} \frac{(1 + a_1 + a_2)\,|E/E_A|^2}{1 + a_1\,|E/E_A| + a_2\,|E/E_A|^2} \quad . \tag{2.16}$$

With (2.13) the behavior of the dielectric permittivity can be written as:

$$\varepsilon_r(E) = \varepsilon_r(0)\,[1 - \tau_\varepsilon(E)]$$
$$= \varepsilon_r(0)\left[1 - \tau_{E_A} \frac{(1 + a_1 + a_2)\,|E/E_A|^2}{1 + a_1\,|E/E_A| + a_2\,|E/E_A|^2}\right] \quad . \tag{2.17}$$

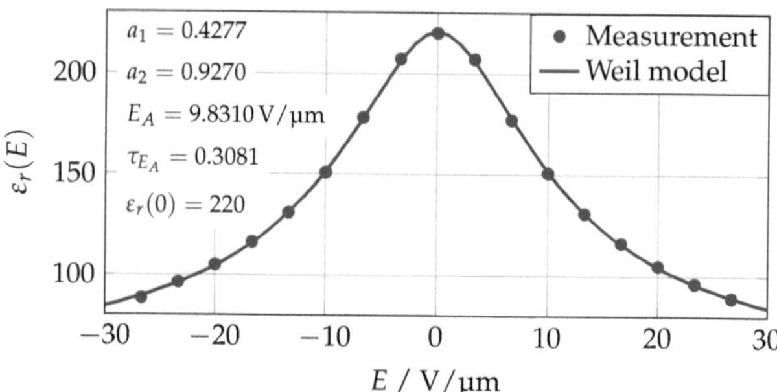

Figure 2.7: Measured ε_r of a Fe-F codoped thick film versus the applied bias field at 5 GHz. Model proposed by Weil [Wei03] accurately predicts the behavior.

With this model, the ferroelectric tunability in the paraelectric phase is completely described by two parameters (a_1 und a_2) and a tunability $\tau_{E_A} = \tau_\varepsilon(E = E_A)$ observed at a tuning field E_A. This model is of specific interest, as it is used in later chapters to investigate and optimize the geometry of tunable components. An extensive comparison between the different models has been performed in [Mau11]. Fig. 2.7 exemplary shows the fitting results obtained with the model for a Fe-F codoped BST layer at 5 GHz.

2.3.2 Process Technology for Tunable Components

The fabrication, of tunable components based on thick film technology can be subdivided into two classes of components:

1. Uniplanar

2. Multilayer

The uniplanar fabrication method is commonly used to structure coplanar waveguides (CPW), interdigital capacitors (IDCs) or meander inductances. These structures can be fabricated after the sintering process of the BST layer. Therefore, the required sintering temperature for the BST layer is of minor importance. The design of multilayered structures intrinsically requires a reduction of BST sintering temperature to the level of low temperature cofired ceramics (LTCC) to avoid infiltration of metal into the pores of the BST layer or even evaporation during firing. The work of

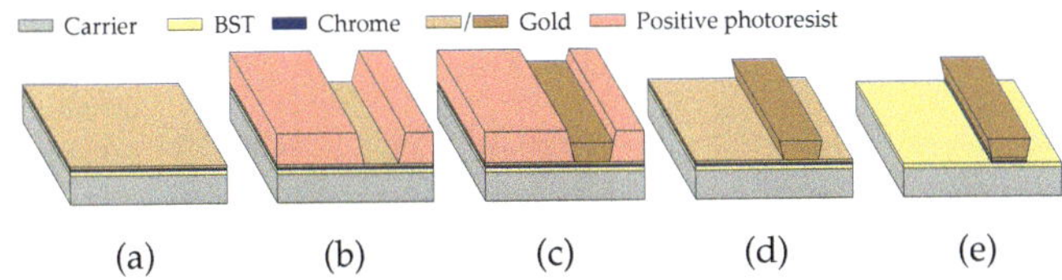

Figure 2.8: Process steps for the fabrication of gold electrodes using a positive photoresist: (a) BST covered carrier substrate with the chrome and gold seed layers, (b) spin coated, UV exposed and developed photoresist, (c) galvanically heightened structures, (d) after removing the photoresist and (e) after removing the seed layers.

Kohler [Koh16] covers the investigation of BST and its adaption to LTCC compatible sintering temperatures.

Uniplanar Structures

The realization of uniplanar components begins with a high temperature compatible carrier substrate, usually alumina (Al_2O_3). For the work here, lapped alumina substrates with a thickness of 650 µm were chosen. The carrier substrate is covered with a BST layer though a (structured) wire mesh (325 wires/in), resulting in a layer thickness in the range of 4 µm to 5 µm. The BST-coated substrate is sintered at 1200 °C in dried air for 10 hours.During firing, the BST layer thickness is reducing to approximately 4 µm.

The patterning process, shown in Fig. 2.8 begins with an evaporation of 60 nm chrome adhesion layer, followed by a 60 nm gold seed layer, see Fig, 2.8(a). The prepared substrate is spin coated with a photoresist, exposed to UV light through a shadow mask and developed, see Fig. 2.8(b). The developed substrate is purged in O_2 plasma for 30 seconds to remove residues of photoresist and developer. The seed layer is galvanically grown to approximately 3 µm through the open areas of the photoresist, see Fig. 2.8(c). Here, AZ 4562 positive (Micro Chemicals) photoresist has been used, which results in a thickness of 6.2 µm, if spin coated at 4000 rpm. After the galvanic bath, the photoresist is removed with acetone and isopropyl alcohol, followed by an O_2 plasma purge for 2 minutes at 200 W RF power. The seed layers are successively removed by selective gold and chrome etchants, see Fig. 2.8(d),(e). The galvanically heightened structures remain. The utilized photo resists, galvanic bath and the etchants for the metals are summarized in Appendix E.

Figure 2.9: (a) Fabricated IDC with 4 fingers on top of a BST layer. (b) The corresponding cross-section with an electrode shape typical for positive photoresist.

Fig. 2.9(a) shows a top view of an IDC varactor with 4 fingers and approximately 3 µm thick and 15 µm wide gold electrodes on top of a 4 µm thick BST layer. The cross-section in Fig. 2.9(b) shows the cone shape of the electrodes, fabricated with a positive photoresist.

Multilayer Structures

The realization of multilayered structures, such as metal insulator metal (MIM) capacitors, first of all requires a low temperature sintered BST material, which was investigated in [Koh16]. The fabrication process itself begins with an LTCC compatible substrate. With a sintering temperature around 800 °C, aluminum nitride (AlN) can be used. It is usually avoided as it reacts with the environment at temperatures around 1000 °C. In analogy to the uniplanar structures, first, an adhesion layer of chrome and gold are evaporated on top of the carrier substrate, see Fig. 2.10(a). The following steps are identical to the procedure of the uniplanar structures. However, the important difference is, that a negative photoresist is used for the lithography. This photoresist produces a slope of the walls, which yields pyramid-like electrodes after the galvanic bath, see Fig. 2.10(b)–(e). This shape reduces the surface tension of the BST paste (or ink) and allows to obtain a homogenous BST layer after the screen or inkjet printing, see Fig. 2.10(f).

Once the bottom electrode is fabricated, it is selectively covered with BST through a prestructured screen and sintered at 850 °C for 1 hour. After the sintering, the top electrode is processed in a similar manner with either positive or negative photo resist, forming a parallel-plate capacitor (MIM).

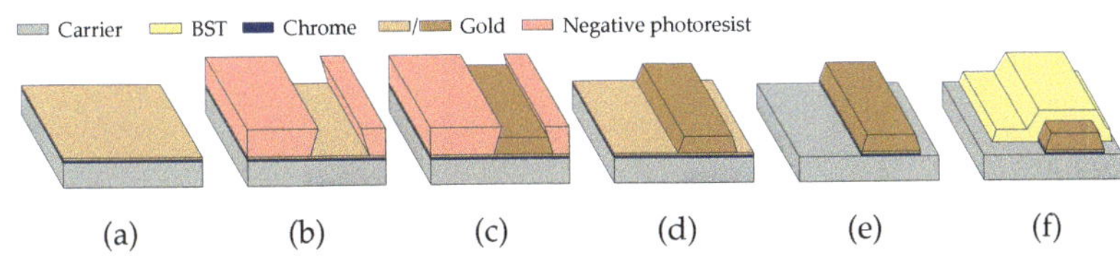

Figure 2.10: Process steps for the fabrication of gold electrodes using a negative photoresist: (a) carrier substrate with the chrome and gold seed layers, (b) spin coated, UV exposed and developed negative photoresist, (c) galvanically heightened structures, (d) after removing the photoresist and (e) after removing the seed layers. Structures covered with a BST layer (f).

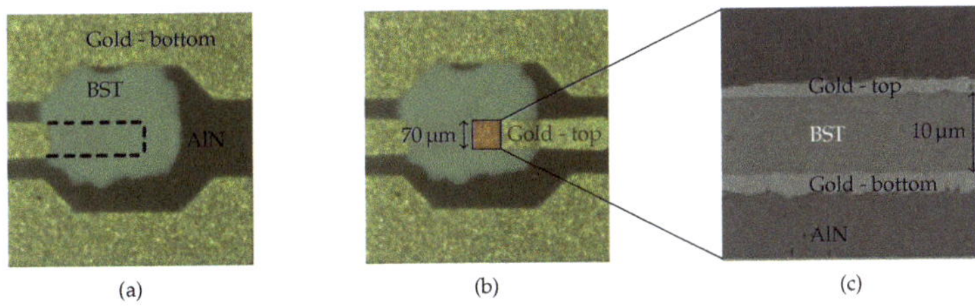

Figure 2.11: (a) Processed gold bottom electrode on top of AlN carrier substrate covered with screen printed BST layer. (b) fabricated top electrode forming a MIM varactor with $70\,\mu m \times 70\,\mu m$ area. (c) Cross-section through the capacitor area.

Fig. 2.11(a) shows fabricated bottom electrode covered with a screen printed BST layer of $10\,\mu m$ after sintering. In a second step the top electrode is processed, forming a parallel-plate capacitor with a $70\,\mu m \times 70\,\mu m$ capacitor area, shown in Fig. 2.11(b). A cross-section of the capacitor is shown in Fig. 2.11(c).

3 Ferroelectric Composite Material and Varactor Optimization

3.1 Multi Dielectric Composites

Ferroelectric materials are constantly under investigation and tailored to specific applications with regard to low effective permittivity ε_r, improved Q or increased tunability τ [ZZY11; Tan+14]. To achieve the desired dielectric property changes, the ferroelectric material is doped, diluted with high Q dielectrics, embedded in a dielectric matrix, stacked in layers or grown as columns [Tan+14]. Therefore, within the last decades, significant effort was put into synthesis and dielectric modeling of such tailored materials.

It is intuitively assumed, that the microwave performance (reduced losses, increased tunability) of the ferroelectrics can be improved beyond limitations, given by the intrinsic properties [Tag+03], by mixing them with non-tunable low-loss dielectrics. This approach was investigated within the DFG granted project (JA 921/37-1). Precise modeling of such engineered materials would allow substantial time saving to design devices with desired properties, since currently most ferroelectric/dielectric composite materials are investigated experimentally only. Mathematically, these composites have been mostly treated for the case of the simplest models, where the components are distributed in layers or columns and poorly describe the nature of real binary composites. Advanced theoretical descriptions have been developed but with constraints on their applicability [She+06].

The *spherical inclusion model* [Ber78] makes assumptions that the electric field is homogeneous throughout the composite, which, as a consequence, is only accurate for small concentrations of additives and becomes severely inaccurate for high dilution ratios. Also, high dielectric contrast between the spherical inclusions and the dielectric matrix leads to significant field re-distribution within the material, making this model unsuitable for ferroelectric/dielectric composites. Still, for small amounts of dilutions, this model accurately describes the nature of composites, and is therefore often used to benchmark more advanced models. Another widely used approach to describe the properties of the composite materials is the *equivalent medium approximation* (EMA) [Bru35]. This method neglects the physical structure of the material and is based on intuitively made statements regarding the effective properties. Nowadays, EMA is widely used to describe the electrical properties of arbitrary linear composites [She+06].

To the best knowledge, currently none of the available models can satisfactorily predict the effective dielectric properties of ferroelectric/dielectric composites, or more general, high dielectric contrast materials. Numeric models have been investigated, but only for small dielectric contrast [KSN00; She+06] or neglect the nature of real composites [Lei15].

In the following, a numerical approach utilizing modern CAD software is employed to extract and model dielectric properties of binary composites. The obtained numeric predictions are compared to the available models. The results are verified with measured BST – magnesium boron oxide (MgBO) composites.

3.1.1 Modeling of Dielectric Properties

Analytical Models

Three different mathematical models allow exact analytical solutions for various concentrations of two or more dielectrics. Fig. 3.1 exemplary shows the commonly used representations starting with (a) columnar model, (b) layered model and (c) spherical inclusion model. The electrodes are assumed at the top and the bottom along the z-axis. The first two models also allow exact analytical solution throughout the entire concentration ratio range, as their geometrical orientation does not alter the fields.

Columnar Model: One possibility to calculate the effective permittivity ε_{eff} and the effective loss tangent $\tan \delta_{\text{eff}}$ of the configuration depicted in Fig. 3.1(a) is the use of energy preservation law:

$$
\begin{aligned}
W_{\text{s,tot}} &= \sum_{i=1}^{N} W_{\text{s,i}} \ , \\
W_{\text{d,tot}} &= \sum_{i=1}^{N} W_{\text{d,i}} \ ,
\end{aligned}
\tag{3.1}
$$

where $W_{\text{s,tot}}$ is the total stored energy and $W_{\text{d,tot}}$ is the total dissipated energy in the equivalent medium, which is constituted by N different materials. The energy

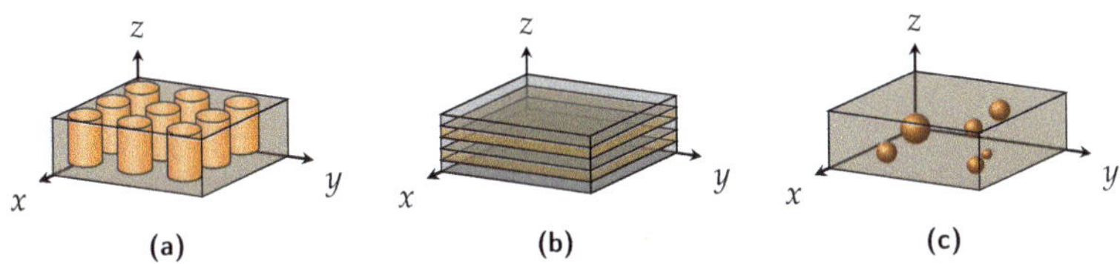

Figure 3.1: Drawing of different composite models. (a) Columnar model (b) Layered model. (c) Spherical inclusion model. Ferroelectric (orange) is embedded into dielectric matrix (gray).

preservation law in (3.1) states, that the total stored and dissipated energy can be written as the sum of stored and dissipated energies of each material independently. Making use of the Pointing theorem one finds the stored W_s and the dissipated energy W_d through the enclosed surface $S = \partial V$ of a volume V [Col00]:

$$W_s = \operatorname{Im} \frac{1}{2} \oint_S \vec{E} \times \vec{H}^* \cdot (-d\vec{S}) = 2\omega \int_V \left(\mu' \frac{\vec{H}\vec{H}^*}{4} - \varepsilon' \frac{\vec{E}\vec{E}^*}{4} \right) dV \ ,$$

$$W_d = \operatorname{Re} \frac{1}{2} \oint_S \vec{E} \times \vec{H}^* \cdot (-d\vec{S}) = \overbrace{\frac{\omega}{2} \int_V \left(\mu'' \vec{H}\vec{H}^* + \varepsilon'' \vec{E}\vec{E}^* \right) dV}^{\text{Polarization damping}} \tag{3.2}$$
$$\underbrace{+ \frac{1}{2} \int_V \sigma \vec{E}\vec{E}^* \, dV}_{\text{Joule heating}} \ .$$

In this work, a binary composite, consisting of a ferroelectric and a dielectric phase is of interest and (3.1) reduces to two summands with the index f and d for ferroelectric and dielectric material. Neglecting the Joule heating term as well as the ferromagnetic contributions μ'' and μ' and assuming $\vec{E} = E\hat{e}_z$, the combination of (3.1) and (3.2) results in:

$$\frac{\omega}{2} \varepsilon'_{\text{eff}} |E|^2 V = \frac{\omega}{2} \varepsilon'_f |E_f|^2 V_f + \frac{\omega}{2} \varepsilon'_d |E_d|^2 V_d \ ,$$
$$\frac{\omega}{2} \varepsilon''_{\text{eff}} |E|^2 V = \frac{\omega}{2} \varepsilon''_f |E_f|^2 V_f + \frac{\omega}{2} \varepsilon''_d |E_d|^2 V_d \ , \tag{3.3}$$

with V_f and V_d being the ferroelectric and the dielectric volume and V being the total volume of the composite material.

The volume fraction coefficient $q = V_f/V$ and $(1-q) = V_d/V$ rewrites (3.3) as:

$$
\begin{aligned}
\varepsilon'_{\text{eff}}(E,q) &= \left(q \cdot \varepsilon'_f(E_f) \, |E_f|^2 + (1-q) \cdot \varepsilon'_d \, |E_d|^2 \right) \frac{1}{|E|^2} \,, \\
\varepsilon''_{\text{eff}}(E,q) &= \left(q \cdot \varepsilon''_f(E_f) \, |E_f|^2 + (1-q) \cdot \varepsilon''_d \, |E_d|^2 \right) \frac{1}{|E|^2} \,.
\end{aligned}
\tag{3.4}
$$

Due to continuity of the electric field $E_{\|,1} \overset{!}{=} E_{\|,2}$, the E-fields in (3.4) can be set to $E = E_f = E_d$ leading to:

$$
\begin{aligned}
\varepsilon'_{\text{eff}}(E,q) &= q \cdot \varepsilon'_f(E) + (1-q) \cdot \varepsilon'_d \,, \\
\varepsilon''_{\text{eff}}(E,q) &= q \cdot \varepsilon''_f(E) + (1-q) \cdot \varepsilon''_d \,,
\end{aligned}
\tag{3.5}
$$

which is a linear combination of both material properties scaling with q. The relation $\tan\delta = \varepsilon''/\varepsilon'$ in combination with (3.5) gives a more convenient representation of the effective losses within the composite:

$$
\tan\delta_{\text{eff}}(E,q) = \frac{q \cdot \tan\delta_f(E)\varepsilon'_f(E) + (1-q) \cdot \tan\delta_d\varepsilon'_d}{q \cdot \varepsilon'_f(E) + (1-q) \cdot \varepsilon'_d} \,.
\tag{3.6}
$$

The tunability $\tau_{\text{eff}}(E,q)$ of the composite material is considered within the E-field dependence of $\varepsilon'_f(E)$ in (3.5). Hence, the effective tunability can be calculated as:

$$
\begin{aligned}
\tau_{\text{eff}}(E,q) &= \frac{\varepsilon'_{\text{eff}}(0,q) - \varepsilon'_{\text{eff}}(E,q)}{\varepsilon'_{\text{eff}}(0,q)} \\
&= \frac{q \cdot (\varepsilon'_f(0) - \varepsilon'_f(E))}{q\varepsilon'_f(0) + (1-q)\varepsilon'_d} \,.
\end{aligned}
\tag{3.7}
$$

Assuming $\varepsilon'_f \gg \varepsilon'_d$, which is a reasonable assumption for the composite treated within this work, (3.5), (3.6) and (3.7) reduce to:

$$
\begin{aligned}
\varepsilon'_{\text{eff}}(E,q) &\approx q \cdot \varepsilon'_f(E) \,, \\
\tan\delta_{\text{eff}}(E) &\approx \tan\delta_f(E) \,, \\
\tau_{\text{eff}}(E) &\approx \tau_f(E) \,.
\end{aligned}
\tag{3.8}
$$

Consequently, as long as the approximation $q\varepsilon'_f \gg (1-q)\varepsilon'_d$ is valid, the tunability and the dielectric losses of the composite are mostly dominated by the ferroelectric material, while the permittivity $\varepsilon'_{\text{eff}}$ linearly increases with q. For the columnar configuration no improvement in dielectric loss or tunability of pure ferroelectric can be achieved, and therefore the overall figure of merit of the material stays constant:

$$\eta_{\text{eff}}(E) \approx \frac{\tau_f(E)}{\tan \delta_f} = \eta_f(E) \; . \tag{3.9}$$

However, this alignment is applicable if tunable substrates with reduced permittivity are required.

Layered Model: For the calculation of layered composites depicted in Fig. 3.1(b) one can again use the energy preservation law and employ (3.4) in accordance to the columnar model. However, since the dielectric layers are now oriented perpendicular to the E-field, field redistribution between the components has to be considered. The derived equations can be found in the Appendix A.

Making the same assumptions as for the columnar model with $\varepsilon'_f \gg \varepsilon'_d$ (A.7), (A.9) and (A.10) reduce to:

$$\varepsilon'_{\text{eff}} \approx \varepsilon'_d \; ,$$
$$\tan \delta_{\text{eff}} \approx \tan \delta_d \; , \tag{3.10}$$
$$\tau_{\text{eff}} \approx \tau_f \cdot 0 \approx 0 \; .$$

This leads to the conclusion, that the properties of the layered composite material are mostly dominated by the dielectric material and the tunability yields 0. Consequently, the layered alignment of the dielectrics achieves a reduction of the material FoM to:

$$\eta_{\text{eff}}(E) \approx 0 \; . \tag{3.11}$$

However, if reduced Curie-Weiss temperature is desired, this material structure can be beneficial[1].

Spherical inclusion model: The spherical inclusion model, considered within this work, consists of a random distribution of ferroelectric spheres in a dielectric matrix as depicted in Fig. 3.1(c). To obtain the effective permittivity $\varepsilon'_{\text{eff}}$ it is assumed, that the ferroelectric spheres are sparsely embedded in the dielectric matrix, such that the electric field can be assumed as homogenous. Therefore the problem can be reduced to the calculation of a single sphere. The integral of the entire volume yields:

$$\frac{1}{V} \int_V \left[D(\vec{r}) - \varepsilon'_d E(\vec{r}) \right] \, \mathrm{d}V \equiv \overline{D} - \varepsilon'_d \overline{E} \; , \tag{3.12}$$

---

[1]For the case of small dielectric concentrations $((1 - q) \ll 1)$, it can be shown that the properties of the composite material correspond to those of a pure ferroelectric material with a reduced Curie-Weiss temperature [She+06].

where the relations $\overline{D} = \frac{1}{V}\int_V D(\vec{r})\,\mathrm{d}V$ and $\overline{E} = \frac{1}{V}\int_V E(\vec{r})\,\mathrm{d}V$ have been used for average displacement $\overline{D}$ and average E-field $\overline{E}$. The field within the dielectric has been assumed as constant, therefore the integrand in (3.12) is zero except within the ferroelectric volume. (3.12) can be rewritten as:

$$\begin{aligned}
\overline{D} - \varepsilon'_d\overline{E} &= \frac{1}{V}\int_{V_f}\left[D(\vec{r}) - \varepsilon'_d E(\vec{r})\right]\mathrm{d}V \\
&= \frac{V_f}{V}\cdot\left(D_f - \varepsilon'_d E_f\right) \\
&= q\cdot\left(\varepsilon'_f E_f - \varepsilon'_d E_f\right) \\
&= q\cdot\left(\varepsilon'_f - \varepsilon'_d\right)E_f \quad,
\end{aligned}$$

(3.13)

where D_f and E_f is the displacement and E-field within the ferroelectric respectively and $q = V_f/V$ being the volume fraction of the ferroelectric. The E-Field E_{sp} within a dielectric sphere with the permittivity ε'_2 embedded in a dielectric with ε'_1 in an external field E is well known [LL63]:

$$E_{\mathrm{sp}} = \frac{3\varepsilon'_1 E}{2\varepsilon'_1 + \varepsilon'_2}.$$

(3.14)

Combining equations (3.14) and (3.13) yields:

$$\begin{aligned}
\overline{D} - \varepsilon'_d\overline{E} &= q\cdot\frac{3\varepsilon'_d\left(\varepsilon'_f - \varepsilon'_d\right)\overline{E}}{2\varepsilon'_d + \varepsilon'_f} \\
\Leftrightarrow\quad \varepsilon'_{\mathrm{eff}}\overline{E} &= \varepsilon'_d\overline{E} + q\cdot\frac{3\varepsilon'_d\left(\varepsilon'_f - \varepsilon'_d\right)\overline{E}}{2\varepsilon'_d + \varepsilon'_f} \quad,
\end{aligned}$$

(3.15)

where the relation $\overline{D} = \varepsilon'_{\mathrm{eff}}\overline{E}$ has been used. Hence, the effective permittivity of the composite is found as:

$$\varepsilon'_{\mathrm{eff}}(q,E) = \varepsilon'_d + q\cdot\frac{3\varepsilon'_d\left(\varepsilon'_f(E) - \varepsilon'_d\right)}{2\varepsilon'_d + \varepsilon'_f(E)}\quad.$$

(3.16)

Note that the field dependence has been added at this point as $\varepsilon'_f(E)$ for convenience reasons. An alternative approach to derive the same equation is given in [YHS93]. However, (3.16) is only valid for small amount of inclusions. Increasing number would inevitably alter the E-field, making previously made assumptions invalid.

Hence, only values for $q < 0.05$ will be treated for this model. Assuming $\varepsilon'_f \gg \varepsilon'_d$ simplifies (3.16) to:

$$\varepsilon'_{\text{eff}}(q) = \varepsilon'_d \cdot (1 + 3q) \ . \tag{3.17}$$

At the same time the tunability $\tau_{\text{eff}}(q,E)$ yields zero. Similar to the layered model, it reduces the material FoM to zero.

To derive the effective loss $\tan \delta_{\text{eff}}$, the dissipated energy within the composite is used. In accordance to (3.3) the effective loss tangent has been derived in Appendix A (A.19) and can be readily found as:

$$\tan \delta_{\text{eff}}(q,E) = \tan \delta_d + \frac{9q\varepsilon'_f(E)\varepsilon'_d \left(\tan \delta_f - \tan \delta_d\right)}{\left(\varepsilon'_f(E) + 2\varepsilon'_d\right)\left(\varepsilon'_f(E) + 2\varepsilon'_d + 3q\left(\varepsilon'_f(E) - \varepsilon'_d\right)\right)} \ . \tag{3.18}$$

In the dilute limit ($q \approx 0$) and in combination with $\varepsilon'_f \gg \varepsilon'_d$ (3.18) reduces to:

$$\tan \delta_{\text{eff}} = \tan \delta_d \ . \tag{3.19}$$

It can be concluded, that in the dilute limit the dielectric properties of the composite are dominated by the dielectric matrix, and therefore no improvement in dielectric properties can be expected for this case. Further, this model only applies for small ferroelectric amounts, and is therefore not suitable to model the entire mixing range from $q = 0$ to $q = 1$.

Equivalent Medium Approximation: (EMA) [Bru35] is the most popular method to model the properties of mixed linear composites with arbitrary concentrations [She+06]. It has been modified and tailored to describe different aspects of composite materials. Here, the modified EMA (MEMA) derived by Sherman et. al. [She+06] is used to model the dielectric properties of ferroelectric/dielectric composites. The approach is found on the variations (perturbation method) of stored δW_s and dissipated energy δW_d, resulting in:

$$\varepsilon'_{\text{eff}}(q,E) = \frac{1}{4}\left[2\varepsilon'_f(E) - \varepsilon'_d - 3(1-q)(\varepsilon'_f(E) - \varepsilon'_d) \cdots \right.$$
$$\left. + \sqrt{8\varepsilon'_d\varepsilon'_f(E) + \left[2\varepsilon'_f(E) - \varepsilon'_d - 3(1-q)(\varepsilon'_f(E) - \varepsilon'_d)\right]^2} \right] \ , \tag{3.20}$$

with q being the ferroelectric fraction of the composite.

The effective quality factor $Q_{\text{eff}} = 1/\tan\delta_{\text{eff}}$ can be written as:

$$\tan\delta_{\text{eff}}(q,E) = \frac{(1-q)c_d^2\varepsilon_d'\tan\delta_d + qc_f^2\varepsilon_f'(E)\tan\delta_f}{\varepsilon_{\text{eff}}'(q,E)\cdot[1-2(1-q)b_d^2-2qb_f^2]} \quad,$$

$$Q_{\text{eff}}(q,E) = \frac{\varepsilon_{\text{eff}}'(q,E)\cdot[1-2(1-q)b_d^2-2qb_f^2]}{(1-q)c_d^2\varepsilon_d'/Q_d + qc_f^2\varepsilon_f'(E)/Q_f} \quad,$$

$$\text{(3.21)}$$

with

$$b_d \equiv b_d(q,E) = \frac{\varepsilon_d' - \varepsilon_{\text{eff}}'(q,E)}{\varepsilon_d' + 2\varepsilon_{\text{eff}}'(q,E)} \quad,$$

$$b_f \equiv b_f(q,E) = \frac{\varepsilon_f'(E) - \varepsilon_{\text{eff}}'(q,E)}{\varepsilon_f'(E) + 2\varepsilon_{\text{eff}}'(q,E)} \quad,$$

$$c_d \equiv c_d(q,E) = \frac{3\varepsilon_{\text{eff}}'(q,E)}{\varepsilon_d' + 2\varepsilon_{\text{eff}}'(q,E)} \quad,$$

$$c_f \equiv c_d(q,E) = \frac{3\varepsilon_{\text{eff}}'(q,E)}{\varepsilon_f'(E) + 2\varepsilon_{\text{eff}}'(q,E)} \quad.$$

Note that in [She+06] the authors relate to q as the dielectric fraction, whereas in (3.20) and (3.21), q is referred to as the ferroelectric fraction. Further, the derived equation $\tan\delta_{\text{eff}} = \cdots$ was rewritten as Q_{eff} with the common relation $Q = \frac{1}{\tan\delta}$. The field dependence of $\varepsilon_f'(E)$ in (3.20) allows no close formula for field dependent tunability $\tau(q,E)$. It is therefore numerically evaluated. The tunability $\tau_{\text{eff}}(q,E)$ can be approximated by changing the relative ferroelectric permittivity of the material $\varepsilon_{r,f}'$.

This is exemplary numerically evaluated for $\varepsilon_{r,d}' = 7$, $\varepsilon_{r,f}'(0) = 1000$, $\varepsilon_{r,f}'(E) = 500$, $Q_d = 2500$, $Q_f = 100$. The results are summarized in Fig. 3.2(a). The extracted tunability as well as the material figure of merit η are shown in Fig. 3.2(b). For the sake of simplicity ε_r' will be from now on referred to as ε without the subscript r for relative permittivity $\varepsilon_r' \equiv \varepsilon$.

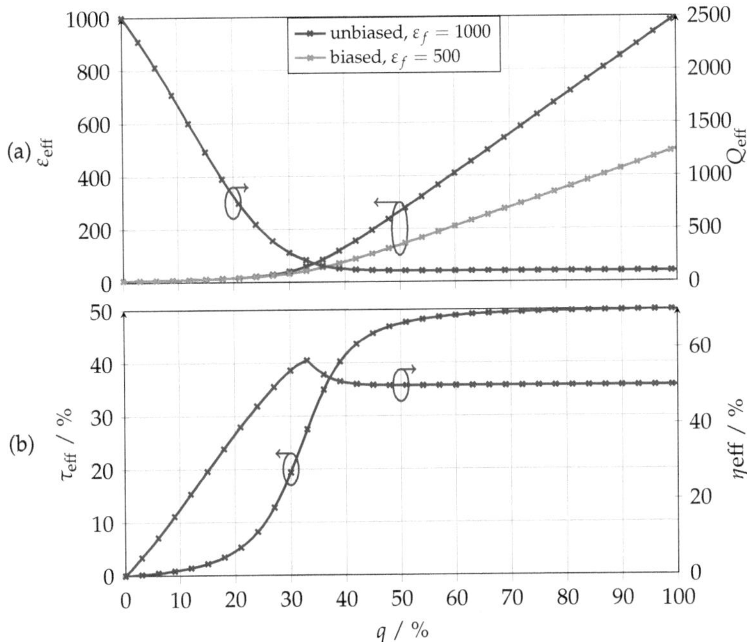

Figure 3.2: Equivalent Medium Approximation (EMA): (a) Effective values of Q and relative permittivity ε_{eff} for biased and unbiased composite material. (b) The extracted tunability τ and material figure of merit η. The following values have been assumed $\varepsilon_d = 7$, $\varepsilon_f(0) = 1000$, $\varepsilon_f(E) = 500$, $Q_d = 2500$, $Q_f = 100$.

This model predicts an increase of the material figure of merit from 50 (pure ferroelectric) to approximately $\eta \approx 60$ around $q = 30\,\%$. However, the effect of field re-distribution is neglected with this approximation and, therefore, it is difficult to make statements about the composite's tunability at this point.

CAD-aided Composite Modeling

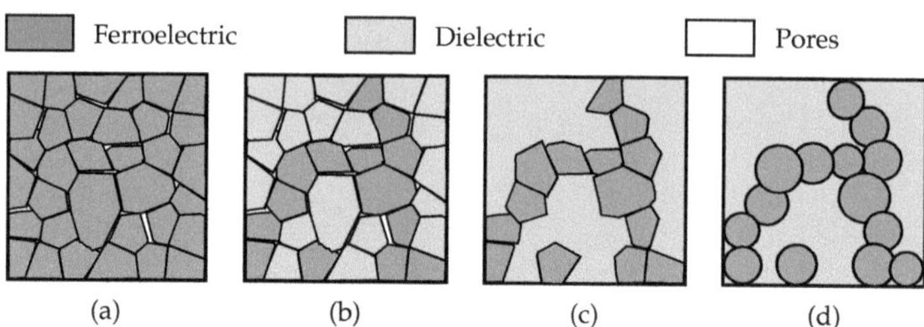

Figure 3.3: (a) Porous ferroelectric ceramic. (b) Porous composite consisting of ferroelectric and dielectric grains. (c) Approximation of the ternary composite [b] as a binary composite, where the pores are treated as dielectric. (d) Approximation of grains with spheres of similar diameter/volume.

Pure, densely sintered ferroelectric ceramics usually consist of interconnected ferroelectric grains with a porosity in the range of 1 % to 5 %. The pores themselves can be considered as a dielectric material with $\varepsilon_p = 1$ and $\tan \delta_p \approx 0$. Therefore, porous single phase ceramic should be considered as a binary composite, see Fig. 3.3(a). Thus, adding a second dielectric to a porous single phase ceramic consequently leads to a ternary ceramic as depicted in Fig. 3.3(b). However, the contrast in dielectric properties of the pores and the dielectric material compared to the contrast between pores/dielectric and ferroelectric is negligible. Therefore, the pores can be modeled as if filled with dielectric material according to Fig. 3.3(c). Hence, instead of treating a ternary composite, one can reduce the complexity to a binary composite. To further simplify the approach, the ferroelectric grains are approximated with spheres of similar diameter or volume as shown in Fig. 3.3(d).

As has been previously mentioned, the *spherical inclusion model* treats the dielectric (ferroelectric) components as isolated spheres and is, therefore, not suitable for large and interconnected number of spheres. EMA neglects the physical distribution of the spheres and relates to the material directly as an effective medium, which does not include the physical distribution of the spheres. Depending on the internal structure, the same ferroelectric volume fraction coefficient $q = V_f/V_{\text{tot}}$, with V_f being the volume of ferroelectric material and V_{tot} being the total volume of the compound can lead to significantly different material properties. Fig. 3.4 depicts some possible configurations, each having the same q but inherently different dielectric properties, which can be also seen as fundamentally different micro

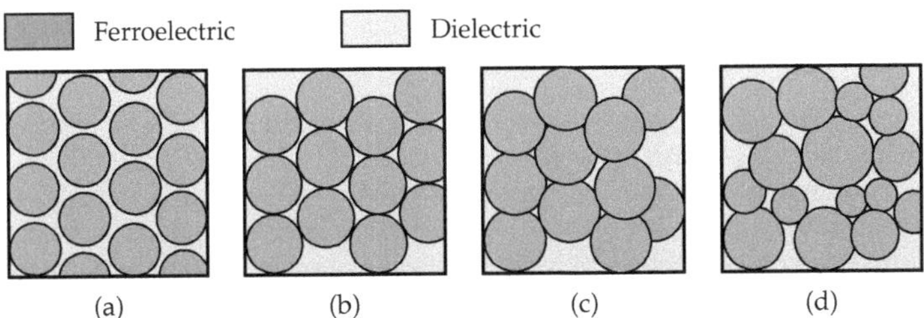

Figure 3.4: A set of possible arrangements with constant ferroelectric amount (a) isolated, (b) touching, (c) overlapping and (d) random.

capacitor networks. Fig. 3.4(a) shows a set of fully isolated ferroelectric spheres and the effective properties are dominated by the dielectric matrix. Fig. 3.4(b) shows a scenario where the ferroelectric spheres touch each other. The influence of the ferroelectric material increases. Fig. 3.4(c) shows a fully connected ferroelectric matrix. In this case, the dielectric properties are dominated by the ferroelectric phase. In Fig. 3.4(d), a random configuration with different sphere radii as well as different sphere interconnections is depicted. This configuration gives the best approximation of a real composite.

In order to accurately predict the dielectric properties of a composite, not only the ferroelectric volume fraction q has to be considered within the model, but also the form of the micro capacitor network, formed by the grains. However, the amount of possible variations is endless such that a large amount of grains has to be considered to obtain statistical average of the expected composite's properties. The advances in numerical methods, as well as the increase in computer performance, have opened up the possibility to perform simulations on a microscopic level, which allow to include the composite's grain configuration. The following assumptions are made for the proposed CAD model:

1. The ferroelectric phase consists of spherical inclusions embedded in a dielectric matrix. The dielectric component is treated as background material (not as individual spheres).

2. The ferroelectric inclusions are are assumed to be isotropic in ε and $\tan \delta$

3. The radii of the spherical inclusions are randomly generated to cover statistical variations and should map the material distribution.

4. The spheres are randomly distributed in space and have adjustable intersection among each other.

To generate a composite model which implements statistical variations of inclusions, the position as well as the diameter of the spheres have to be randomized. The random number generator of a computer is equally distributed between 0 and 1, which is unsuitable for the problem. Also, placing the spheres randomly (but equally distributed) in space is inefficient: with increasing number of spheres the available space for new spheres reduces, and hence, the chance to generate an occupied, and therefore, forbidden position increases. Even if taken into consideration, the placement of new spheres in a densely populated volume is challenging. The whole problem is therefore split in two steps: first, generation of statistically distributed sphere radii and subsequent generation of a coordinate map to densely pack the spheres.

Random Grain Size Distribution

The grain size distribution of a sintered poly crystalline ceramic follows a log-normal distribution [Men69], with a probability density function (PDF):

$$\text{PDF} \equiv p(x) = \begin{cases} \frac{1}{\sqrt{2\pi}\sigma x} \exp\left(-\frac{(\ln(x)-\mu)^2}{2\sigma^2}\right), & x > 0 \\ 0, & x \leq 0 \end{cases} . \tag{3.22}$$

The PDF can be used to obtain the cumulative distribution function (CDF) [Pre+07]:

$$\begin{aligned} \text{CDF} \equiv \int_0^x p(x')dx' &= \frac{1}{2}\left(1 + \text{erf}\left(-\frac{\ln(x)-\mu}{\sqrt{2}\sigma}\right)\right) \\ &= \frac{1}{2}\text{erfc}\left(-\frac{\ln(x)-\mu}{\sigma}\right), \end{aligned} \tag{3.23}$$

where $\text{erf}(x)$ is the error function and $\text{erfc}(x) = 1 - \text{erf}(x)$ is the complementary error function. The inverse of the CDF (y-axis $\mapsto$ x-axis) can be used to map the equal distribution $[0,1]$ to the log-normal distribution. Consequently, the inverse of the CDF (y-axis $\mapsto$ x-axis) can be used to map any equally distributed random number $x \in [0,1]$ to a measured distribution according to the inverse CDF:

$$\begin{aligned} d(x) \equiv \text{CDF}^{-1} &= \exp\left(\mu - \sqrt{2}\sigma \cdot \text{erfc}^{-1}(2x)\right) \\ &= \exp\left(\mu + \sqrt{2}\sigma \cdot \text{erf}^{-1}(-1+2x)\right). \end{aligned} \tag{3.24}$$

Generation of Dense Sphere Packings

The dense alignment of uniform spheres has been mathematically treated and exact solutions can be found for numerous orientations [CS99]. In three dimensions, uniform spheres can be arranged as dense as:

$$\Omega = \frac{\pi}{3\sqrt{2}} \approx 0.74, \tag{3.25}$$

with Ω being the volume fraction of spheres normalized to the total volume V. For more complicated packing problems (e.g. multi radii), no exact solution is known, such that heuristic optimization algorithms have to be used in order to find at least quasi-optimum configurations. The approach followed in this work was motivated by [MSS09], where 2-dimensional multi radii hard discs are densely packed employing laws of condensed matter physics. The authors equip each disc with physical behavior similar to that of gas particles and cool the gas such that the particles condense (including attractive force between the discs), forming a dense disc packing. In this work, the approach was simplified and adapted to three dimensions. A set of N equally and randomly distributed spheres with random diameter d are generated at time $t_0 \equiv t = 0$ in a rigid container with zero kinetic and angular velocity, hence only having potential energy. The sphere properties are set to:

- Radius: randomly generated according to CDF^{-1}

- Mass: $m \propto r^3$

- Hardness: fully rigid

- Damping: after each collision half of the kinetic energy is absorbed

- Angular damping: after each collision half of the angular energy is absorbed

- Friction: coefficient set to 0.5

The system is exposed to gravity $g = 9.8\,\text{m/s}^2$ and calculated for $t > 0$ with $\Delta t = 1\,\text{ms}$ until the kinetic energy of the system reaches a minimum. The physics engine from *Bullet Physics* [Bul] was used for the calculation of collisions. When the simulation starts at t_0, the spheres fall until they collide with the container bottom, as illustrated in Fig. 3.5(a). After time $t \gg t_0$ the system comes to rest. The result is a dense multi radii sphere packing (within the accuracy of simulation), which is illustrates in Fig. 3.5(b). The maximum achieved density is in the range of $\Omega = 0.68$, which is lower than (3.25) due to cavities, which form during the simulation. An improvement can be achieved by introducing attractive force between the spheres as

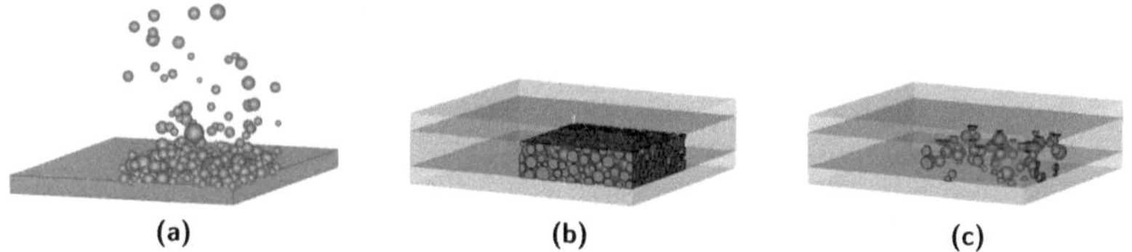

Figure 3.5: (a) Situation of spheres at the start of the simulation. (b) Densely packed, truncated ferroelectric grains in a dielectric matrix forming a parallel-plate capacitor. (c) Reduced ferroelectric amount by reducing the number of spheres.

has been described in [MSS09] or e.g. shaking the configuration to avoid formation of vacant areas. For the work here, the achieved value is sufficient. Fig. 3.5(c) shows a configuration with a reduced amount of spheres. The densely packed sphere configuration (coordinates and diameter) is exported in a text file for further processing with a full-wave EM-simulation.

Electromagnetic Simulation

A list of sphere radii and coordinates is imported into CST Studio Suite between two perfect conductor electrodes, forming a parallel-plate capacitor according to Fig. 3.5(b). The list contains randomly arranged spheres, touching each other in a single point (within the accuracy of the approach described in Sec. 3.1.1). The sphere configuration is simulated with a *frequency domain solver*, extracting key properties such as ε_{eff} and $Q_{\text{eff}} = 1/\tan\delta_{\text{eff}}$. To achieve sufficient simulation accuracy in combination with curved elements (spheres), a tetrahedral mesh was chosen. A compromise between mesh density and calculation speed was made with 50000 tet / μm^3.

As a consequence of the generation approach described in Section 3.1.1, the imported list is unsorted in radii and randomized in coordinates. Therefore, to sweep the ferroelectric amount of the composite material only $0 < n \leq N$ entries of the list are imported, while the remaining $N - n$ entries are omitted, see Fig. 3.6(a). Thus, the same configuration with a reduced ferroelectric amount is obtained and re-calculated. The volume V_f of the imported spheres is used to calculate the ferroelectric fraction of the composite $q = V_f/V$. Fig. 3.7 exemplary shows the obtained effective properties versus q at 100 MHz of a composite, assuming $\varepsilon_f = 2600$, $Q_f = 100$, $\varepsilon_d = 7$, $Q_d = 2500$, which are typical values for tunable ferroelectrics and high-Q dielectrics. The actual field-dependent tunability of the ferroelectric $\tau_f(E)$ is

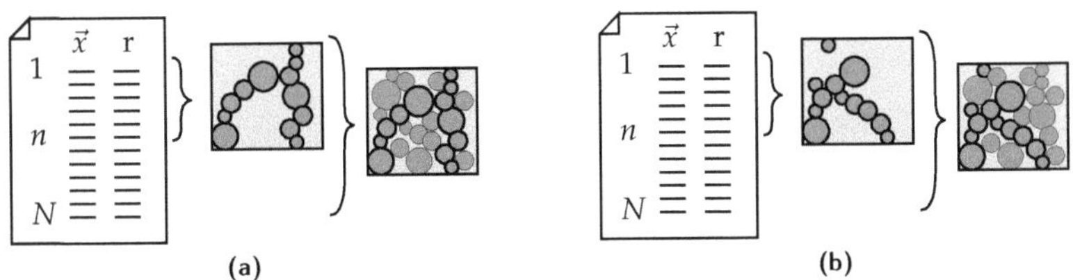

Figure 3.6: (a) Illustration of the generated list with sphere coordinates which is imported in CST Studio Suite. (b) Different permutation of the same list (entries change their position). To sweep the ferroelectric amount of the composite, the list is not fully imported. The import of n entries leads to different sphere configurations.

approximated by a reduction of the ferroelectric permittivity, e.g. from $\varepsilon_f = 2600$ to $\varepsilon_f = 2000$ giving $\tau = 23\,\%$, followed by a re-calculation with the same geometry.

As has been mentioned in Section 3.1.1, the spheres are generated such that they touch each other in a single point. By slightly scaling the radii of the generated spheres by a factor S an additional degree of freedom is introduced for the model. This allows the study of sphere interconnection influence, which has been postulated by [Gie+08]. To illustrate the influence of the sphere interconnection in Fig. 3.7, the radii were scaled by $\pm 0.5\,\%$.

However, the available memory, and hence the amount of mesh cells, limits the simulated amount of spheres and the corresponding volume. Here, a volume of approximately $25\,\mu\text{m}^3$ with a maximum number of $N = 800$ spheres (to achieve $q = 68\,\%$) was used.

For q values around the percolation threshold, a single sphere can make a difference between an interconnected ferroelectric matrix and isolated sub-matrices. Even though a large statistical sample is evaluated, the formation of the spheres dominates the dielectric properties around the percolation threshold. This issue is illustrated in Fig. 3.6. It shows two permutations of the same sphere configuration (the order of the spheres in the list is permuted), where exemplary only n spheres are imported, leading to different configurations. Importing the whole list results in the same configuration. Fig. 3.8 shows simulated statistical distribution of the effective permittivity ε_{eff} for two fixed ferroelectric volume fractions $q = 10\,\%$ and $q = 20\,\%$. A value below the percolation threshold and a value around the percolation threshold, respectively.

Two major aspects can be observed from the graphs:

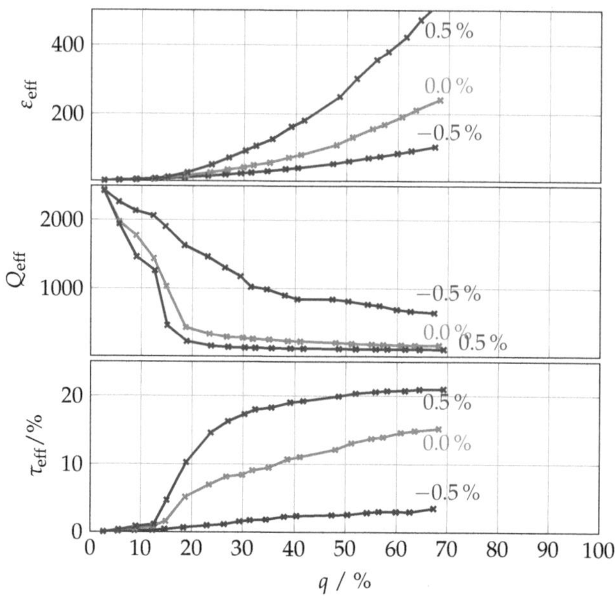

Figure 3.7: Influence of the grain scaling S on the effective permittivity ε_{eff}, the effective quality factor Q_{eff} and the effective tunability τ_{eff}.

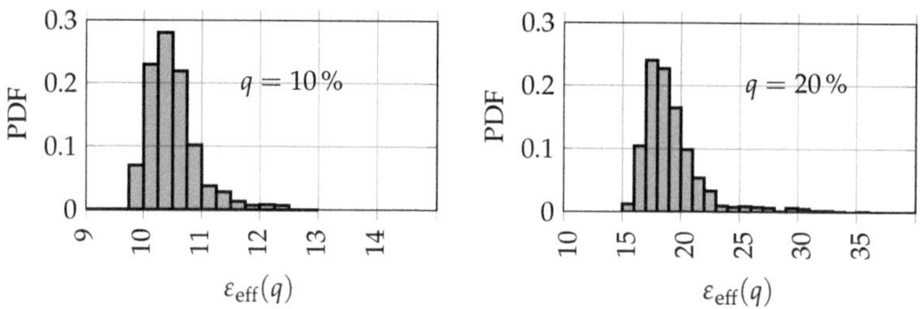

Figure 3.8: Distribution of ε_{eff} obtained with the simulation for $q = 10\,\%$ and $q = 20\,\%$. Each graph contains statistics of 600 simulations.

1. The distribution function of the effective permittivity follows a log-normal distribution, which appears to originate from the log-normal distribution of the sphere diameters. Karkkainen et al. [KSN00] obtain a Gaussian shaped curve for homogenous spheres.

2. The values below the percolation threshold have a significantly lower deviation than the value around the percolation, which is plausible, since in this region the chance of an interconnected sphere network is lower than for ferroelectric fraction around the percolation threshold.

To address this issue and to improve the statistics of the simulation, the sphere list is permuted and recalculated multiple times to obtain a mean value with the corresponding deviation.

3.1.2 Processing and Characterization

To test the proposed modeling approach, (BST) was used as the ferroelectric and magnesium boron oxide (MBO) as the dielectric material:

$$q \cdot Ba_{0.6}Sr_{0.4}TiO_3 + (1-q) \cdot Mg_3B_2O_6 \ , \tag{3.26}$$

with q changed from 0 to 1 in steps of 0.1.

The synthesis of BST powder was done, using a modified sol-gel process as described by Zhou et. al [Zho+10]. The ceramic powder was obtained by calcining the sol-gel-derived BST precursor under purified dried air at a temperature of 900 °C. MBO was synthesized separately via mixing and ball milling of MgO and H_3BO_3 according to the stoichiometric ratio of 3:2 and subsequent calcination at 1300 °C for 2 h. After that, the derived BST and MBO powders were mixed and ball milled in alcohol and zirconia grinding media for 4 hours in a planetary ball mill to obtain the different compositions. To ensure an optimized homogenization of each powder phase, these suspensions were additionally homogenized in a laboratory stirred media mill. After spray drying of these suspensions, the granulated powders were uniaxially pressed into 10.8 mm wide and 1.5 mm thick cylindrical pellets and sintered at 1100 °C for 2 hours. Due to sintering shrinkage, the resulting discs have a width of 8.4 mm to 8.6 mm. The morphology of the samples was determined by scanning electron microscopy (SEM) (Zeiss Supra 55). In order to evaluate the grain size distribution of BST and MBO, the SEM-images were processed with analySIS pro (Olympus). Fig. 3.9(a) shows an obtained image, while Fig. 3.9(b) shows color-coded grain diameters. The diameters d of at least 400 grains from each sample

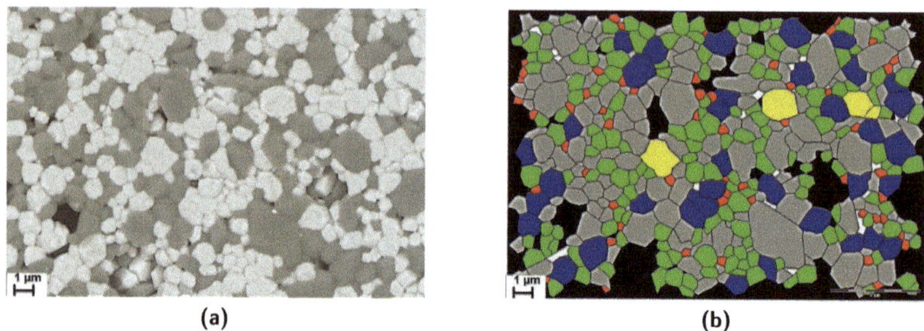

(a) **(b)**

Figure 3.9: (a) Extraction of grain diameters of a sintered BST ceramic from a SEM image. (b) Color-grouped grains to visualize grains of the same diameter.

were analyzed to obtain the particle size distributions. Fig. 3.10 shows the obtained distribution for $q = 50\%$ composite, fitted with (3.23) to obtain the parameters $\mu = 0.043\,15(1663)$ and $\sigma = 0.427\,84(1377)$. It also shows the CDF plotted with the obtained parameters μ and σ to calculate CDF^{-1}, which is required for the generation of random sphere diameters. Subsequently, the samples were lapped to a thickness of $0.5\,\mathrm{mm}$ using alumina lapping powder.

Top and bottom electrodes with diameters of $3\,\mathrm{mm}$ and $5\,\mathrm{mm}$, respectively, were screen-printed on both sides of the lapped discs with the silver-platinum paste

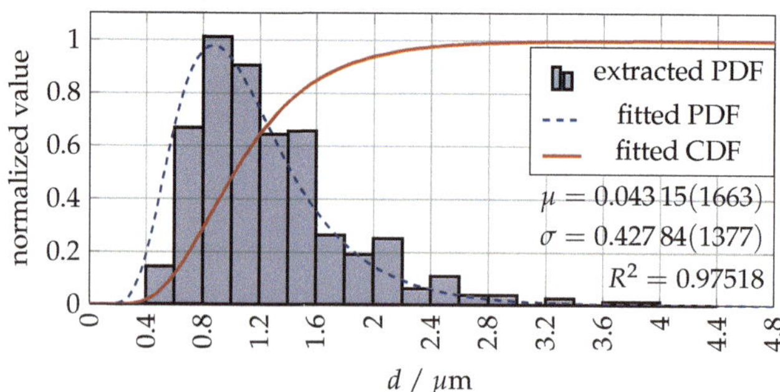

Figure 3.10: Statistical evaluation of the graphically obtained grain diameters for random diameter generation.

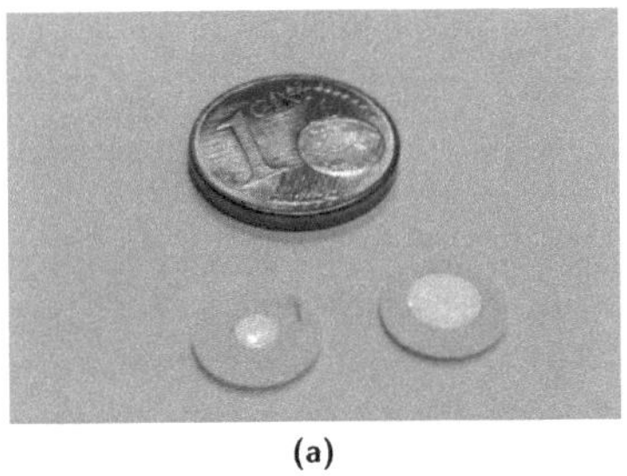
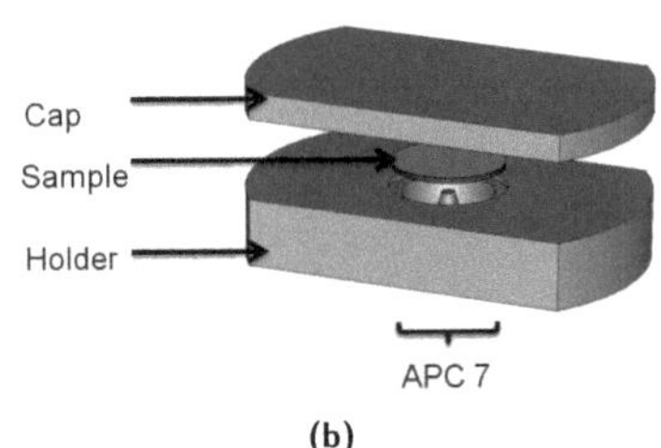

(a)   (b)

Figure 3.11: (a) 0.5 mm thick and $\approx$ 8.5 mm wide samples for impedance spectroscopy. (b) Test fixture for impedance spectroscopy, utilizing the APC7 connector standard.

C 1076 SD (Heraeus GmbH) and fired for 10 minutes at 850 °C. The asymmetry between the top and bottom electrodes allows keeping an exact overlap area (and therefore capacitance) of the parallel-plate capacitor. This has the advantage that the total capacitive area is dominated only by the smaller electrode. Consequently, the misalignment between the top and the bottom electrode has a reduced effect, which allows more accurate permittivity extraction.

The discs were characterized in a fixture with bias voltages up to $1.1\,\text{kV} \equiv 2.2\,\text{V}/\mu\text{m}$ to study the tunability of the composite material. The fabricated test fixture as well as the dielectric samples are depicted in Fig. 3.11. The implemented characterization setup utilizes an APC7 connector standard to allow measurements with the impedance analyzer (Keysight E4991B). The bias voltage was applied through a bias-tee (Picosecond Model 5531). The capacitance C and Q were measured at room temperature from 1 MHz to 100 MHz. The fabricated capacitors do not fulfill the condition of $A \gg d$ to neglect the fringing fields, and hence, do not allow the calculation with $\varepsilon_{\text{eff}}(U_B) = C(U)d/(A\varepsilon_0)$. Therefore, the formula derived by Chew and Kong [CK80] was used to extract ε_{eff} of a capacitor with circular electrode radius r and an electrode distance d:

$$C \approx \frac{\pi r^2 \varepsilon_{\text{eff}} \varepsilon_0}{d} \left\{ 1 + \frac{2d}{\pi r \varepsilon_{\text{eff}}} \left[\ln \frac{r}{2d} + (1.41\varepsilon_{\text{eff}} + 1.77) \right. \right.$$
$$\left. \left. + \frac{d}{r}(0.268\varepsilon_{\text{eff}} + 1.65) \right] \right\} . \tag{3.27}$$

The results obtained by (3.27) were verified by full-wave EM simulation, since it was originally derived for identical top and bottom electrodes. The measured data was evaluated at 100 MHz and the obtained results are summarized in Table 3.1.

Table 3.1: Dielectric properties of the composite materials at room temperature at 100 MHz. The tunability was measured at 1.1 kV bias voltage.

Constituents		Analysis		Dielectric properties					
BST	MBO	Por.	q	$\bar{\varepsilon}$	$\Delta\bar{\varepsilon}$	Q	ΔQ	$\bar{\tau}$	$\Delta\bar{\tau}$
vol.%	vol.%	vol.%	%	-	-	-	-	%	%
10	90	7.0	9.3	14.6	0.1	1500	707	0.7	0.2
20	80	7.6	18.5	27.2	1.0	338	45	3.0	0.1
30	70	8.1	27.6	45.5	0.2	197	35	5.8	0.4
40	60	8.2	36.7	84.4	0.8	166	1	8.9	0.3
50	50	7.2	46.4	161.6	16.9	218	19	14.3	0.5
60	40	9.1	54.5	298.5	0.8	168	2	15.5	0.3
70	30	11.0	62.3	442.0	76.1	165	5	14.7	1.9
80	20	7.7	73.8	675.2	50.7	116	6	17.3	1.1
90	10	4.4	86.0	884.5	0.1	93	15	22.5	0.7
100	-	29.88	70.12	1250	Hakki-Coleman setup				

In addition, a sample with $q = 100\,\%$ was fabricated to determine the properties of pure BST. The sintering temperature was kept at 1100 °C for compatibility reasons with MBO composites. After sintering, the sample showed a diameter of 10.7 mm and a thickness of 5 mm with a porosity of 29.88 %, which can be attributed to the relatively low sintering temperature. The characterization was performed in accordance to the method described by Hakki-Coleman [HC60] and Courtney [Cou70] with a measured resonance frequency of the TE_{011}-mode at 1.15 GHz. Knowing the dimensions and the resonance frequency, the permittivity can be calculated according to:

$$\varepsilon = 1 + \left(\frac{c}{\pi D f}\right)^2 (\alpha_1^2 + \beta_1^2) \,, \tag{3.28}$$

with

$$\alpha_1^2 = \left(\frac{\pi D}{\lambda}\right)^2 \left[\varepsilon - \left(\frac{\lambda}{2H}\right)^2\right] \,,$$

$$\beta^2 = \left(\frac{\pi D}{\lambda}\right)^2 \left[\left(\frac{\lambda}{2H}\right)^2 - 1\right] \,,$$

where H and D are the diameter and height of the sample and λ is the free-space wavelength. This gives a sample's relative permittivity of $\varepsilon_f = 1250$.

As has been previously mentioned, this porous material can be treated as a binary component, therefore the actual ferroelectric permittivity was approximated with the formula derived by Bruggeman [Bru35] (EMA), yielding $\varepsilon_f \approx 2250$ (dense BST ceramic), which is in good agreement with the literature data [Jeo04]. The pores have been assumed with $\varepsilon_d = 1$.

3.1.3 Comparison of Models and Measurement Results

The proposed CAD-assisted modeling approach is compared to the analytical models introduced in Section 3.1.1. Fig. 3.12 summarizes the results. Both, the layered and the columnar models strongly deviate from the measurement data. The results of the MEMA, provided with the dielectric properties of pure BST ($\varepsilon_f = 2250$) and pure MBO ($\varepsilon_d = 7$), shows strong deviation above $q > 30\,\%$. It can be fitted (MEMA$_{\text{FIT}}$) to match the measured permittivity of the composite materials as shown in Fig. 3.12. However, its fitting parameters, see Table 3.2, do not correspond to the reported values of pure $Ba_{0.6}S_{0.4}TiO_3$ at room temperature ($\varepsilon \in [2000 - 4000]$) [Jeo04] and do not correspond to the measured value of $\varepsilon_f = 2250$, see Table 3.1. The CAD model, with its additional degree of freedom for sphere interconnection ($S = 1.011$), accurately models the effective permittivity of the measured composites. Table 3.2 summarizes the obtained fitting parameters and the CAD-model parameters.

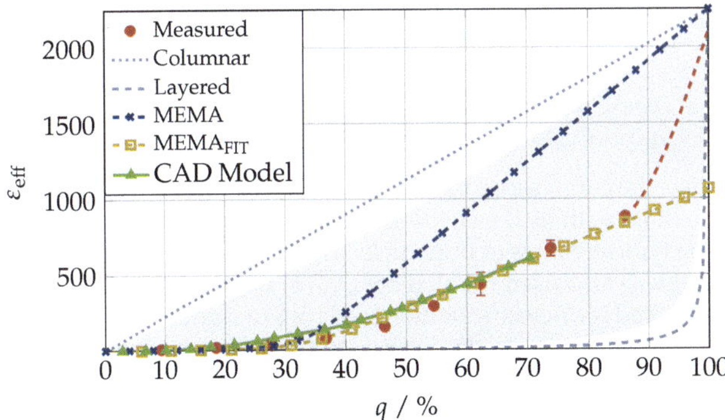

Figure 3.12: Comparison of effective permittivity of the models. The shaded area denotes the Hashin-Shtrikman bounds.

Table 3.2: Extracted dielectric properties and the deviation σ between model and the measured data for Q_{eff} and $\varepsilon_{\mathrm{eff}}$.

	ε_f	ε_d	σ_ε	Q_f	Q_d	σ_Q
Layered	2250	7	-	100	2500	-
Columnar	2250	7	-	100	2500	-
Spherical	2250	7	-	100	2500	-
MEMA	2250	7	3,07	100	2500	36
MEMA$_{\mathrm{FIT}}$	1069	6	0.27	14.5	7586.5	0.59
CAD ($S = 1.011$)	2250	7	0.28	100	2500	0.11

Both models (MEMA and CAD), as well as the measured data are well inside the Hashin-Shtrikman bounds [HS62], which are described in three dimensions by:

$$\varepsilon_{\mathrm{eff,min}} = \varepsilon_d + \frac{q}{\frac{1}{\varepsilon_f - \varepsilon_d} + \frac{1-q}{3\varepsilon_d}} \ ,$$

$$\varepsilon_{\mathrm{eff,max}} = \varepsilon_f + \frac{1-q}{\frac{1}{\varepsilon_d - \varepsilon_f} + \frac{q}{3\varepsilon_f}} \ . \tag{3.29}$$

However, the large dielectric contrast leads to broadened bounds, leaving large space for variations, see Fig. 3.12. To compare the results a deviation metric is defined in accordance to least mean squares as:

$$\sigma = \frac{1}{N} \sum_{i=1}^{N} \left(\frac{y_i - y_m}{y_m} \right)^2 , \tag{3.30}$$

with y_i and y_m being the y-values of the models and the measurement at the same x-coordinates, respectively. The obtained values are summarized in Table 3.2.

Fig. 3.13 shows the comparison of the models for the effective quality factor. Here, Q_{eff} was also fitted with (3.21), yielding fitted parameters for Q_f and Q_d, which do not match the typical values of pure dielectric and pure ferroelectric, see Table 3.2. Further, (3.21) was also used to calculate $Q_{\mathrm{eff}}(q)$ with measured ($\varepsilon_d = 7$ and $\varepsilon_f = 2250$) dielectric and ferroelectric values, leading to a significant mismatch between the measurements and the model for $q < 40\,\%$, see Fig. 3.13. However, the CAD-model accurately maps the trend of the measurement data. Around the percolation threshold, the error bars of the model are increased, since the variations of the statistical sample have large impact on the micro-capacitor network and a small change can lead to strong changes in Q_{eff}. Also, measurement of high-Q values with the impedance spectroscopy is challenging, which is also indicated in the large

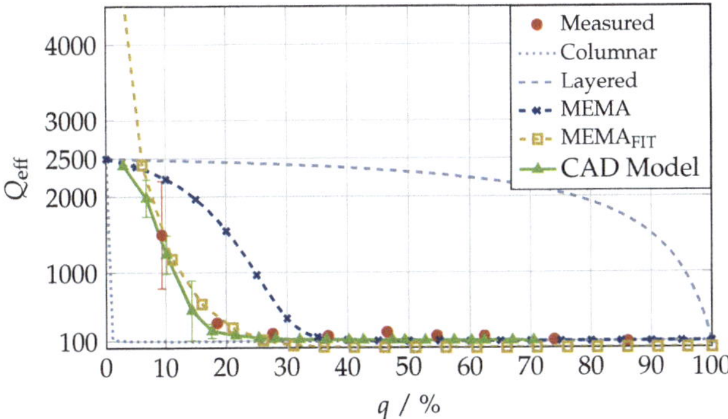

Figure 3.13: Comparison of effective quality factor Q_{eff} of the models. The CAD approach maps the measured data with good accuracy.

error bar for low q/high-Q values in the measured data. This makes it difficult to assess the accuracy of the CAD model for the lower q values. The same metric, as for the effective permittivity (3.30), was used to compare the models and the measured data, see Table 3.2, proving good agreement between the CAD model and the measurement. However, a closer look at the measurement data around $q = 45\,\%$ reveals a local maximum, which has been observed in various publications with BST and MBO [ZZY11; Zha+10]. This effect cannot be explained by the properties of the constituents and the laws of electrodynamics and requires further investigation. Currently, this behavior is expected to arise from formation of interface layers or a reaction between BST and MBO. This result shows that even the extensive modeling approach, presented in this work, which is considering more effects than [Bru35; She+06; Zha+10], cannot fully describe the nature of real composites.

To estimate the tunability in MEMA, columnar and layered model, the permittivity ε_{eff} was calculated for two different ferroelectric permittivies $\varepsilon_f(\text{untuned}) = 2250$ and $\varepsilon_f(\text{tuned}) = 1800$, which corresponds to a ferroelectric tunability $\tau_f = 20\,\%$. A similar procedure was performed for the CAD model. Here, the permittivity of the ferroelectric spheres was reduced to 1800 followed by a recalculation. The results are depicted in Fig. 3.14. The measured tunability shows a roughly linear increase over q, while both MEMA and the CAD model predict a strong increase around the percolation at $q = 20\,\%$ and $q = 10\,\%$, respectively. The deviation can be explained by the simple assumptions made for simulation and calculation of the tunability. The redistribution of the electric field, which is responsible for the tunability in the

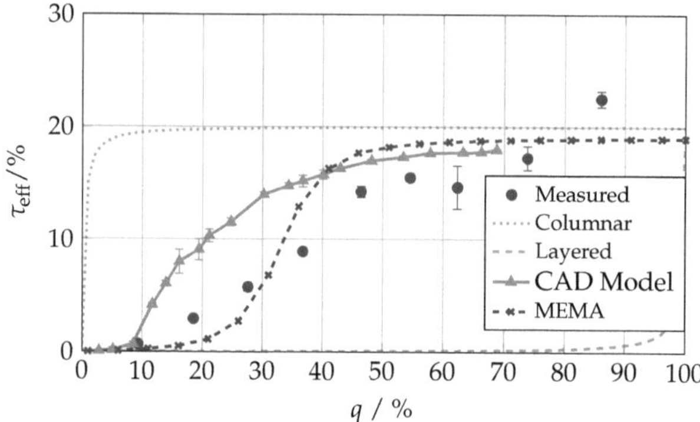

Figure 3.14: Comparison of the models with regard to the effective tunability. None of the models can accurately describe the measured tunability.

measurement was not included in the simulation and is also not included by MEMA. For small ferroelectric concentrations, the field in the ferroelectric component is reduced as a consequence of the permittivity contrast. Consequently, the tunability in the measurement does not show a strong increase as predicted by both models. To include the field redistribution effect, the fields within the ferroelectric sphere have to be considered.

The results show that the equivalent medium approach does not sufficiently predict the dielectric properties over the entire mixing range. Though it can be fitted to describe the measured dielectric properties, its fitted parameters do not accurately describe the properties of pure MBO and pure BST. Additional degrees of freedom are required to consider the grain interconnection, as well as to allow variable percolation threshold. The numerical investigation shows, that the effective dielectric properties of composites can be described as a combination of the dielectric properties of both constituents, which combine to an effective value according to the laws of electrodynamics.

However, while effective permittivity $\varepsilon_{\mathrm{eff}}$ follows the models throughout the entire mixing range, the effective quality factor Q_{eff} cannot be fully explained by a combination of the constituents' properties. A closer look shows an unexpected increase in Q_{eff}, which can be only attributed to morphology, interface effects or reaction between the constituents. This requires further analysis of the composites.

Due to the simple assumption, which neglects the field redistribution, the tunability of the composites was not modeled satisfactory within this work. To improve the accuracy, one should first calculate the field within the ferroelectric inclusions and in a second step replace them with individual spheres having a permittivity corresponding to the field within the spheres. Here, the spheres have been replaced neglecting the field.

3.2 Varactor Topologies: Design and Optimization

The design and optimization of tunable RF components is complicated and involves numerous different parameters, which are influencing each other. The following list gives only a sample of these variables.

1. Material properties

   - Quality factor Q

   - Permittivity ε

   - Doping

   - Thermal stability

2. Material processing

   - Bulk, thick- or thin film

   - Milling

   - Sintering temperature

3. Component geometry

   - Planar or multilayer

   - Gaps, size and thickness of layers

   - Tunability

   - Self-resonance

4. Circuitry related

   - Response time

   - Operation frequency

Hence, the optimization cannot be carried out in a single step for all properties. This section focuses on the layout and design optimization regarding component geometry and circuitry related properties of tunable, ferroelectric varactors, especially interdigital capacitors (IDCs).

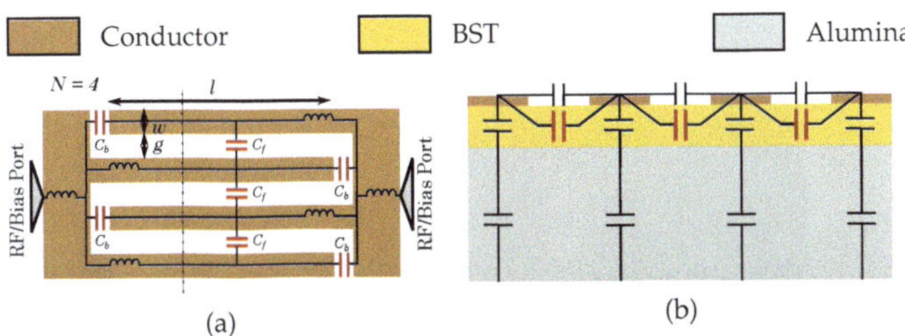

Figure 3.15: (a) Topview on an IDC varactor on top of a BST layer with desired (red) capacitance and parasitic (black) inductors. (b) Cross section along the dashed line with desired (red) and parasitic (black) capacitors.

3.2.1 Uniplanar Thick Film Varactors

The geometry of an IDC dictates the properties of the final component, especially regarding achievable capacitance values, resulting quality factor, self-resonance frequency and tunability. Fig. 3.15 shows a top and cross-section view with an overlay of the resulting reactive elements. It becomes apparent that trying to design an IDC varactor, the desired characteristics (red capacitors in Fig. 3.15) inevitably influence the parasitic elements, which have a significant impact on the component's properties. This issue has been previously addressed utilizing methods of conformal mapping [Gev+96] but with a reported deviation between full-wave simulations and calculation of up to 40 % at 6 GHz [Mau11]. Therefore, ultimately CAD methods are indispensable to accurately predict the characteristics of the components.

In this work, Keysight's Advanced Design Studio (ADS) software, which is utilizing the method of moments (MoM) is used to model parametrized varactors. The advantage of this approach, compared to previous work on CAD modeling of varactors [Gie09; Sch07a] is the direct applicability in a circuit simulator. ADS allows a layout-based EM-simulation of the component, which can be directly used in a circuit. Varactor models similar to those in Fig. 3.15 were implemented with a full set of parameters including number N, length l, gap g and width w of the IDC fingers. Further, the layer structure of the substrate as well as the height of conductive layers can be easily modified to a specific application.

The implemented model allows to sweep the parameters in order to find a layout with the aimed capacitance including parasitic effects. It also allows to draw conclusions on the dependence of the aimed capacitance value $C(N,l,w,g,...)$ and other

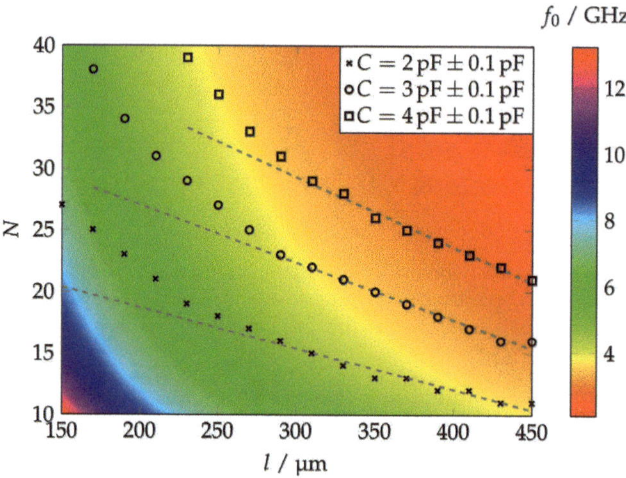

Figure 3.16: Dependence of self-resonance frequency f_0 over the length l and number N of the IDC fingers. Iso-contours give capacitance values around $C = 2\,\mathrm{pF}$, $3\,\mathrm{pF}$ and $4\,\mathrm{pF}$.

variables e.g. the corresponding resonance frequency $f_0(N,l,w,g,...)$. Fig. 3.16 shows the dependence of f_0 versus N and l and allows conclusions about the self-resonance of an IDC. A higher number of short IDC fingers yields a higher self-resonance frequency. Further, a linear dependence of l between N can be observed for $l > 300\,\mathrm{µm}$. In this region the finger capacitors (C_f in Fig. 3.15(a)) dominate. For finger length below $300\,\mathrm{µm}$ the influence of the finger tip towards the RF/bias electrode (C_b in Fig. 3.15(a)) is increasing, which results in a deviation from the linear dependence between l and N for a fix capacitance value.

As has been mentioned in Section 2.3, the BST layer reduces the permittivity when exposed to an electric field. Hence, applying a bias voltage between the IDC fingers reduces the permittivity and, as a result, the capacitance. The layout shown in Fig. 3.15(a) is impractical since one of the IDC electrodes would have to be on high potential (several hundred volt), which would require additional bias decoupling, when deploying the varactor into a circuit. Fig. 3.17 shows two common IDC topologies with intrinsically decoupled bias electrodes [Sch07a]. A significant difference between both models can be seen regarding the RF and bias field distribution which

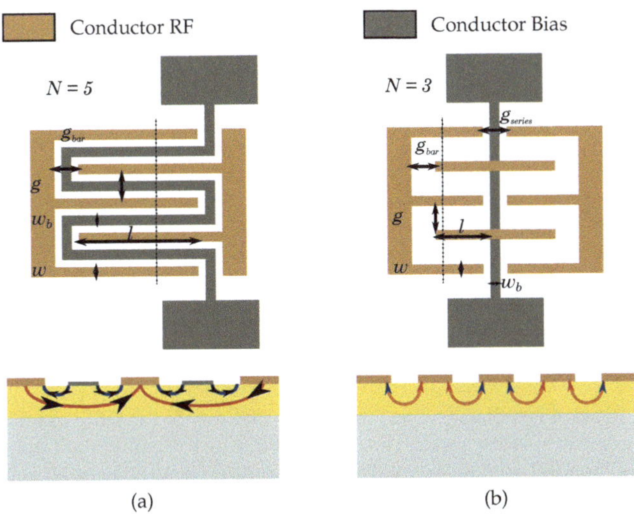

Figure 3.17: (a) IDC varactor with a resistive bias electrode between the fingers. (b) Double (series) IDC varactor with resistive middle electrode. Both topologies allow for decoupling of DC and RF. The fields are color-coded with red (RF) and blue (bias).

is depicted by red (RF) and blue (bias) lines. In Fig. 3.17(b) both the RF field and the bias field overlap leading to an expected higher tunability compared to Fig. 3.17(a). The proposed topologies implement highly resistive In_2O_3:SnO_3 (84:16 wt.%) (ITO) bias electrodes with a thickness between 10 nm and 50 nm to prevent leakage of RF into the bias path [Sch07a]. However, simulations and measurements have shown, that the resistivity of the bias electrode has a significant influence on the overall characteristics of the component. Other materials such as nickel or chromium were used [Saz13] but with similar results.

These topologies were previously investigated with time consuming finite differences time domain (FDTD) [Sch07a], where only parts of the whole component have been modeled and simulated. Fig. 3.18 shows the impact of the resistivity and width of the bias electrodes on the overall varactor C and Q with the presented ADS model, where the whole structure has been used for the simulation. For the RF electrodes the conductivity of gold ($\sigma = 4.1 \cdot 10^7$ S/m) was assumed. To include dielectric

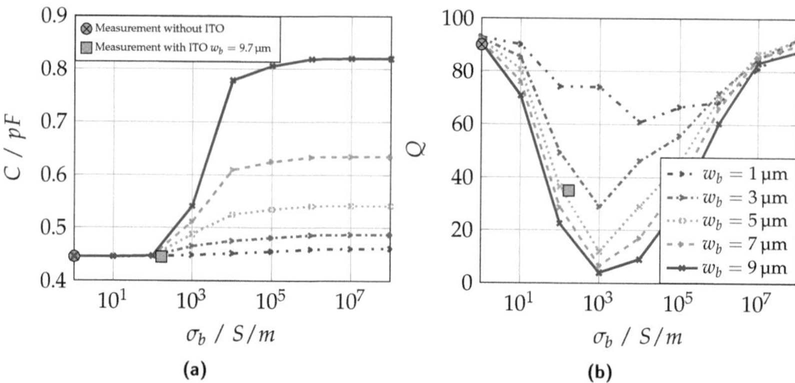

Figure 3.18: (a) Capacitance C and (b) quality factor Q of the IDC varactor versus resistivity σ_b and width w_b of the bias electrode. The measured data points were taken from [Sch07a] with an approximate ITO conductivity of $\sigma_b \approx 170\,\text{S/m}$.

losses, a BST layer quality factor $Q_{\text{BST}} = 100$ was assumed at $2\,\text{GHz}$. The dielectric losses of the carrier substrate were neglected. The result of this investigation is, that the resistivity of bias electrodes cannot be sufficiently reduced in order to maintain the quality factor of the component.

However, beside the fact that the resistive electrodes deteriorate the varactor's properties, it is also limiting in terms of biasing and power handling capability. Insufficient RF-choke, as a consequence of insufficient resistance, leads to a destruction of the bias electrodes when operated at high power, see Fig. 3.19. Therefore, the resistive bias lines were replaced by gold electrodes and external components are used to implement RF/DC decoupling. Guidelines for the choice will be given in Section 3.2.3.

To include tunability of the components, ADS allows to create numerous substrates, each with an individual layer profile and dielectric properties. This is used to approximate tunability for the proposed model. A set of substrates with varying BST layer permittivity was created and an IDC layout with a certain geometry is calculated for each individual substrate. The results are stored in an internal ADS database for each configuration and can be retrieved from the database for the use in the circuit simulator. Fig. 3.20 illustrates the approach. It has to be noted at this point that the recently released version of ADS2016 allows to implement

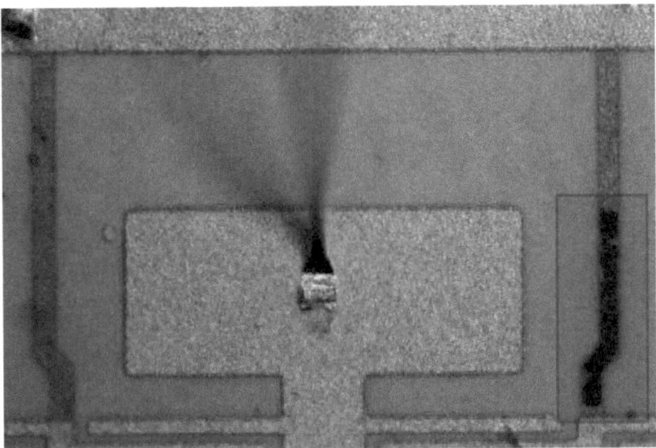

Figure 3.19: Shaded area shows the RF power induced destruction of the resistive chromium lines.

the substrate tunability in a more convenient way, avoiding creation of numerous substrates. However, when the models were developed, no direct influence on a substrate layer's properties were possible and the presented work-around was performed to include tunability. Assuming a tunability of $\tau = 50\,\%$ a set of identical substrates with BST layer permittivity ranging from $\varepsilon_r = 200$ to $\varepsilon_r = 100$ is created. However, this method neglects the actual field dependent tunability of the substrate layer and, therefore, empirical values for the minimum permittivity are used. Further, this approach homogeneously reduces the permittivity of the substrate and does not allow the calculation of non-linear effects when e.g. operating the IDC capacitor under high RF power. Still, the proposed approach allows fast and flexible synthesis of tunable circuits with sufficient accuracy, which will be demonstrated Section 5.3.

To investigate the actual field-dependent tunability of IDC varactors, a model was implemented in CST Studio Suite. The same layer structure as for the ADS model was implemented with a parametrized IDC geometry on top of a layered substrate, which correspond to the case where no bias voltage is applied. An RF simulation is performed to extract properties in the untuned state. Next, an electrostatic simulation is performed, followed by the evaluation of the DC field amplitude within each mesh-cell of the non tunable BST layer. Depending on the field amplitude, the permittivity $\varepsilon_{BST}(E)$ is calculated with the analytical model described in Section 2.3.1 (2.17). The fitting parameters of a thick film, obtained in Section 2.3.1 are used to describe the permittivity for any given field strength within the BST layer ($a_1 = 0.4277$, $a_2 = 0.9270$, $E_A = 9.8310\,\mathrm{V/\mu m}$, $\tau_{E_A} = 0.3081$).

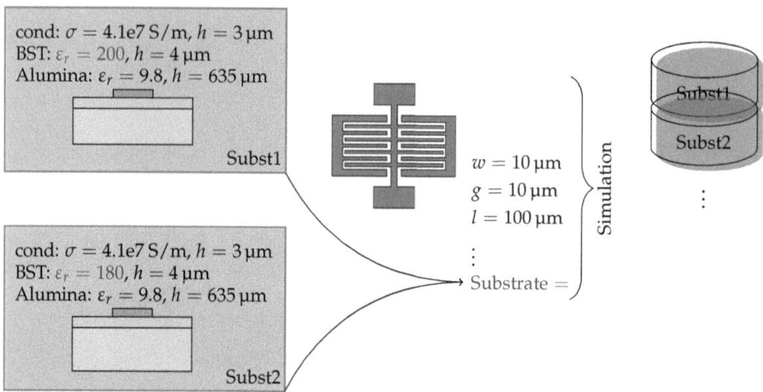

Figure 3.20: Implementation of tunable capacitors in ADS. An IDC geometry is simulated for a set of substrates with varying BST layer permittivity. The results are stored in a database and can be used in circuit simulation.

Next, the non-tunable BST layer is replaced one mesh cell after the other with substrate bricks that have the permittivity calculated from the model. Iterations between the DC simulation and brick replacement can increase the accuracy but is usually kept at one iteration. Finally, an RF simulation is performed and key parameters, such as effective permittivity $\varepsilon_{\mathrm{eff}}$, tuned capacitance, effective tunability τ_{eff} etc. can be extracted.

Fig. 3.21 summarizes the model approach. For visualization, the tuned substrate bricks are colored from black (barely tuned) to red (tuned). By varying the geometry of the IDC and BST layer their interdependence was studied. The dependence between the BST layer thickness h, IDC finger gaps g as well as the BST layer permittivity $\varepsilon_{\mathrm{BST}}$ on the tunability was studied. The BST film thickness was varied between $1\,\mu\mathrm{m}$ to $10\,\mu\mathrm{m}$ in $1\,\mu\mathrm{m}$ steps. The finger gaps were swept from $5\,\mu\mathrm{m}$ to $50\,\mu\mathrm{m}$ in $5\,\mu\mathrm{m}$ steps. The finger width and conductor height were kept constant at $15\,\mu\mathrm{m}$ and $3\,\mu\mathrm{m}$, respectively.

This investigation was carried out for BST layer permittivities $\varepsilon_{\mathrm{BST}} = 200$ and $\varepsilon_{\mathrm{BST}} = 400$ to investigate the impact of the permittivity and to find the optimum finger spacing and achieve maximum tunability of the IDC varactors. The bias voltage was adjusted to keep constant bias field of $10\,\mathrm{V}/\mu\mathrm{m}$.

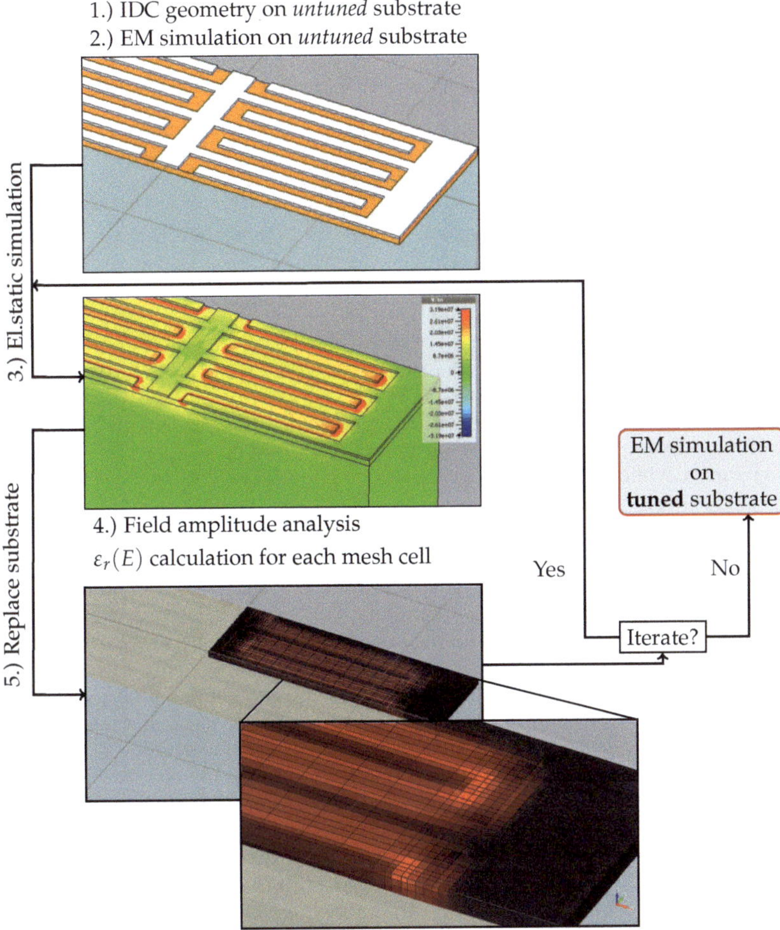

Figure 3.21: Flow diagram of modeling of tunable substrate in combination with IDC varactors in CST Studio Suite. A non-tunable substrate is replaced by bricks with permittivity calculated in accordance to (2.17) and the E-field amplitude in each cell.

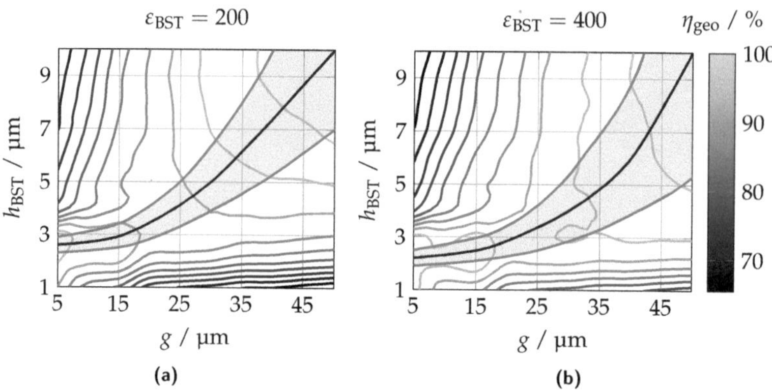

Figure 3.22: Extracted geometrical tuning efficiency η_{geo} with the BST layer permittivity (a) $\varepsilon_{BST} = 200$ and (b) $\varepsilon_{BST} = 400$. The solid curve gives the extracted maximum, while the shaded areas are denoting the uncertainty of the simulation.

Fig. 3.22 shows the obtained results. The solid line shows the maximum value, while the shaded area gives the uncertainty. For large finger gap g, the obtained results are in good agreement with the investigations of [Gie09] and [Mau11] with an optimum geometrical tuning efficiency $\eta_{geo} = \tau_{c,max}/\tau_{m,max}$ for a gap g to BST layer height h_{BST} ratio of $\approx 5 : 1$. They defined the geometrical tuning efficiency η_{geo} as a ratio of the achieved tunability of a component $\tau_{c,max}$ and a maximum material tunability $\tau_{m,max}$. However, as the values become smaller, the optimal ratio changes to approximately 2:1. This is reasonable as the contribution of the nontunable finger capacitance through air is increasing. However, to achieve the required bias field, a compromise between the maximum geometrical efficiency and the maximum tolerable bias voltage have to be made. A finger gap of $15\,\mu m$ has been chosen within this work. The obtained results show, that the developed models allow extensive optimization and modeling of tunable, planar structures on tunable substrates, which so far has only been possible for a fraction of the components.

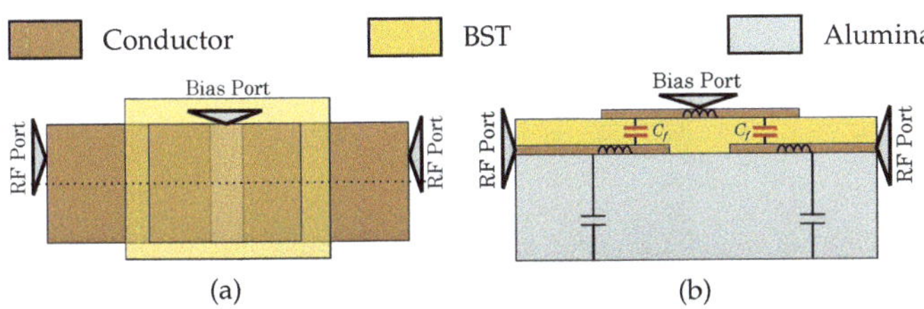

Figure 3.23: (a) Top view of a double MIM varactor. (b) Cross section of the double MIM varactor with major parasitics.

3.2.2 Multilayer Thick- and Thin Film Varactors

Thick Film Ferroelectric Layers

The MIM topology has significantly less interdependent variables, and therefore, does not require extensive modeling as in the case of the IDC capacitors. Nevertheless, an ADS model was implemented to accurately model the inductive and capacitive parasitics. The layer structure was changed to two conductive layers between BST, forming a metal isolator metal (MIM), parallel plate capacitor on top of a carrier substrate, as shown in Fig. 3.23. The tunability of the components was implemented similar to the IDC model in ADS by using substrates with different BST layer permittivity. Compared to IDCs, this topology leads to significantly better approximation of tunability for real components, as the whole BST layer is homogeneously exposed to the bias field.

Thin Film

The thin film varactor geometry does not significantly vary from the layout of the presented MIM thick film varactors. Due to the fabrication process of the BST layer (usually PLD), the isolator thickness can be reduced to several 100 nm. The has a number of advantages and disadvantages:

- Tunability: higher film quality, in combination with reduced thickness, enables high tunability values $\geq 70\,\%$ at lower bias voltage

- Tuning voltages: lower BST thickness $\Rightarrow$ higher tuning fields at lower voltage

- Lower linearity as a consequence of low bias fields

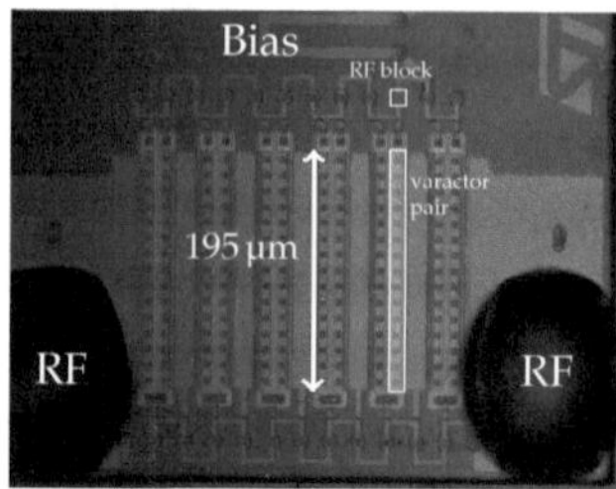

Figure 3.24: Micro photograph of a 1.2 pF thin film varactor from STMicroeletronics. RF and bias electrodes are highlighted. Note that at least one of the RF electrodes has to provide reference potential for the biasing.

- Reduced power handling capability due to smaller size of varactors as a consequence of small electrode distance.
- Acoustic resonances

The acoustic resonance is an intrinsic problem of high quality thin film BST layers and can barely be avoided. In this, case the acoustic resonance is in the range of 4 GHz, see Appendix B. To increase the linearity, as well as to improve the power handling capability, the thin film varactors have to be serialized. Fig. 3.24 shows the micro photograph of a commercial 1.2 pF thin film varactor from STMicroelectronics. To improve the linearity and the power handling capability, 24 varactors with an individual size of $8 \times 195\mu m$ are serialized. A single varactor pair is highlighted in gray. Note, that in this case a varactor pairs are structured vertically in multiple layers, with a total of 12 planar varactor pairs. This on the other hand sets additional demands on the DC/RF decoupling, which, due to space limitation and number of elements can only be implemented as resistors. One decoupling area is highlighted in the image. Note that at least one of the RF electrodes has to provide reference potential for the biasing. However, the resistive decoupling itself reduces the tuning speed, which will be the focus of the following investigations.

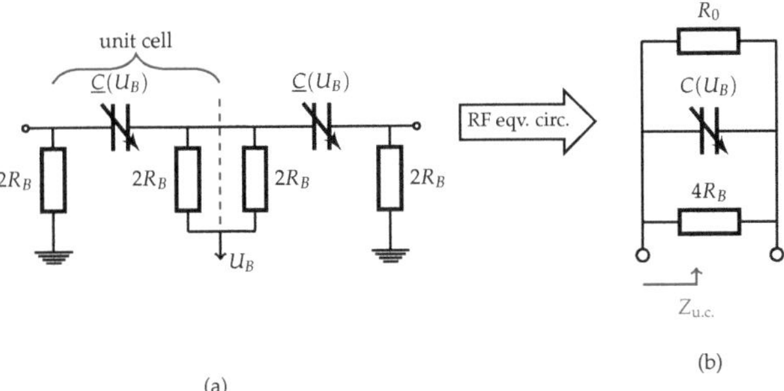

Figure 3.25: (a) Equivalent circuit of resistively decoupled double varactor. (b) Unit cell representation to calculate the impact of R_B on the RF performance on the varactor.

3.2.3 External Biasing Concepts: Resistive and Inductive Decoupling

Resistive Decoupling

Aiming for linear devices at high RF power an even number of series varactors is beneficial. It avoids DC potential on the RF electrodes, while increasing the total linearity and power handling capability, compared to a single capacitance cell. Fig. 3.25 shows the equivalent circuit of a double varactor with two series lossy capacitors $\underline{C}(U_B)$. Shunt resistors R_B are placed as RF choke. In the following, the dimensions of R_B are evaluated. The RF impedance $Z_{u.c.}$ of a single unit cell (Fig. 3.25(b)) can be written as:

$$Z_{u.c.} = \frac{j4Q_0 R_B}{j\omega C(U_B)R_B + Q_0(j - 4\omega C(U_B)R_B)} \; . \tag{3.31}$$

The application of external bias voltage leads to an increase of Q and to a reduction of C [Gev09]. Therefore, it is plausible to carry out the investigation for the minimum capacitance $C_{min} \equiv C(U_B = U_{max})$ and a maximum $Q_{max} \equiv Q(U_B = U_{max})$ to dimension the bias resistor R_B. Equation (3.31) rewrites to:

$$Z_{u.c.} = \frac{j4Q_{max}R_B}{j\omega C_{min}R_B + Q_{max}(j - 4\omega C_{min}R_B)} \; , \tag{3.32}$$

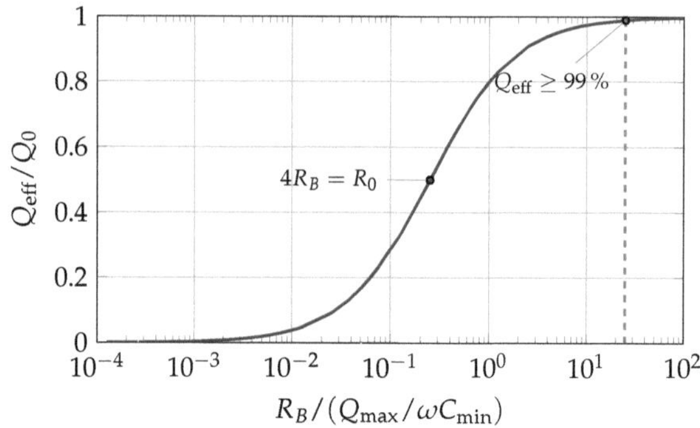

Figure 3.26: Impact of the bias resistor R_B on the effective quality factor of a unit cell.

and can be written as a complex capacitance

$$\underline{C}_{\text{u.c.}} = \frac{1}{j\omega Z_{\text{u.c.}}} = C_{\text{min}} - j\frac{4\omega C_{\text{min}}R_B + Q_{\text{max}}}{4\omega Q_{\text{max}}R_B} \quad , \tag{3.33}$$

with Q_{max} being the the maximum quality factor of the lossy capacitor $\underline{C}(U_B)$ given as $Q_{\text{max}} = \omega C_{\text{min}}R_0$. Hence, the effective capacitance and quality factor are given as:

$$C_{\text{eff}} = \Re\{\underline{C}_{\text{u.c.}}\} = C_{\text{min}} \quad ,$$
$$Q_{\text{eff}} = -\frac{\Re\{\underline{C}_{\text{u.c.}}\}}{\Im\{\underline{C}_{\text{u.c.}}\}} = Q_{\text{max}}\frac{4\omega C_{\text{min}}R_B}{Q_{\text{max}} + 4\omega C_{\text{min}}R_B} \quad . \tag{3.34}$$

Consequently, the capacitance of a unit cell is unaffected by the bias resistors. However, the effective quality factor Q_{eff} strongly depends on the bias resistor R_B since both R_0 and R_B are in parallel and the smaller of both dominates.

Fig. 3.26 shows Q_{eff} normalized to Q_{max} versus $R_B/R_0 \equiv R_B/(Q_{\text{max}}/\omega C_{\text{min}})$. To maintain 90 % to 99 % of Q_{max}, the bias resistor should be:

$$2.5 \cdot Q_{\text{max}}/\omega C_{\text{min}} \leq R_B \leq 25 \cdot Q_{\text{max}}/\omega C_{\text{min}} \quad . \tag{3.35}$$

Consequently, the time constant $\tau_{95\,\%} = 3RC$ for approximately 95 % of bias voltage across the varactor yields:

$$\tau_{95\,\%} = 3 \cdot 4R_B C(U_B) \quad . \tag{3.36}$$

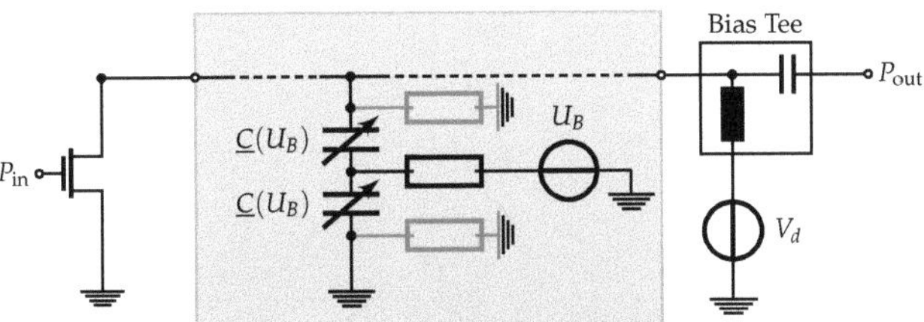

Figure 3.27: Effect on a double varactor placed between a transistor and a bias-tee. The outer (gray) resistors seen by the bias voltage source are shorted.

Considering that the charging of the capacitor starts with C_{max} gives:

$$30\frac{Q_{max}C_{max}}{\omega C_{min}} \leq \tau_{95\%} \leq 300\frac{Q_{max}C_{max}}{\omega C_{min}} \ . \tag{3.37}$$

Assuming capacitance properties $C_{max} = 2\,\text{pF}$, $C_{min} = 1\,\text{pF}$, $Q_{max} = 100$ and $\omega = 2\pi \cdot 2\,\text{GHz}$, yields $R_B \in [20..200]\text{k}\Omega$ and as a consequence $\tau_{95\%} \in [0.48..4.8]\mu\text{s}$. Aiming for high power applications it is instructive to work with the upper limits of R_B, as RF power leakage can also destroy the resistors similar to resistive bias lines in Fig. 3.19. In terms of tuning speed, this value is sufficient for static tuning applications or dynamics below $210\,\text{kHz}$.

However, the the derived equations are only valid for the case, where the varactor is treated isolated. Placing a resistively decoupled varactor in a circuit can short circuit the outer resistors seen by the bias voltage source as shown in Fig. 3.27. In this case, the resistance seen by the bias voltage source reduces from $4R_b$ to $2R_b$. Consequently the time constant is reducing by a factor 2. Further, having one of the outer electrodes on a high reference potential leads to asymmetric biasing curves of the series varactors, which has been investigated in [Mau11]. Here, the focus is put on the time response of the series varactors.

To verify the predicted response time, a tunable capacitor was implemented as a Taylor series with only even contributions $C(U_B) = C_0 + a_2 U_B^2 + a_4 U_B^4 + a_6 U_B^6 \cdots$ to obtain a symmetric response for positive and negative bias voltages. The coefficients were extracted by fitting a measured MIM varactor with a tunability of $\approx 50\%$ at $200\,\text{V}$, see Fig. 3.28(a). Fig. 3.28(b) shows the time domain simulated voltage across a unit cell of a tunable $C(U_B)$ and non-tunable capacitor C_0 treated isolated and with short circuited outer bias resistors R_B. It shows, that the non-tunable capacitors have a slower response time, which is reasonable, as the capacitance is constant

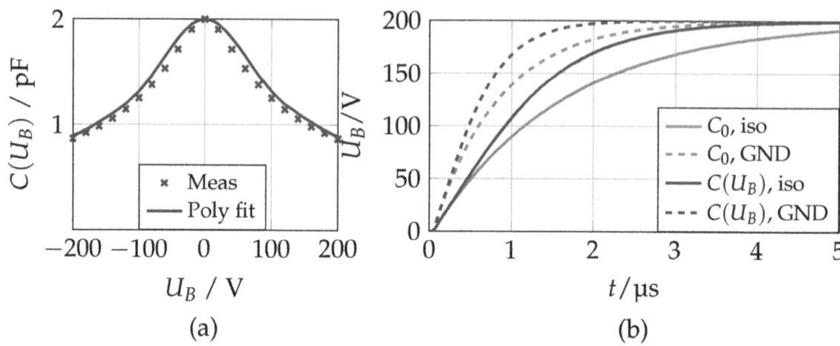

Figure 3.28: (a) Polynomial fit of a measured capacitance $C(U_B)$. (b) Simulated voltage across a unit cell of a non-tunable (C_0) and tunable ($C(U_B)$) capacitor with common and isolated bias and RF ground.

throughout the charging process. The simulation also shows, that putting varactors in a circuit can influence the response time by a factor of 2.

The derived equations (3.35) and (3.37) (which include strong simplifications of tunability) are in good agreement with the non-linear time domain simulation. It can be concluded that for most applications, the response time of a resistively decoupled tunable capacitor is sufficient and is in the range of µs, when aiming for RF frequencies around 2 GHz.

Another consequence is, that when cascading unit-cells e.g. quad varactor, the outer cells are responding faster than the inner cells. When applying bias voltage, the outer capacitors charge faster and, hence, have a smaller capacitance values, compared to the inner unit cells. As a consequence, the outer cells would have to sustain higher power levels, until the inner cells have completely charged and reach their final capacitance value. Fig. 3.29 shows the equivalent circuit of a quad varactor, placed between two bias tees of an RF port. In accordance to previous calculations, the outer cells (marked blue) are responding twice as fast as the inner cells (marked red).

$$P_{\text{cell}} = \frac{C_o(U_B)C_i(U_B)}{C_{\text{cell}}(U_B) \cdot (2C_i(U_B) + (N-2)C_o(U_B))} \cdot P_0 \; , \qquad (3.38)$$

where $C_{\text{cell}}(U_B)$ is either the outer cell $C_o(U_B)$ or the innter cell $C_i(U_B)$ and P_0 is the total power. In a limit, where the outer cells reduced its capacitance by a factor of 2

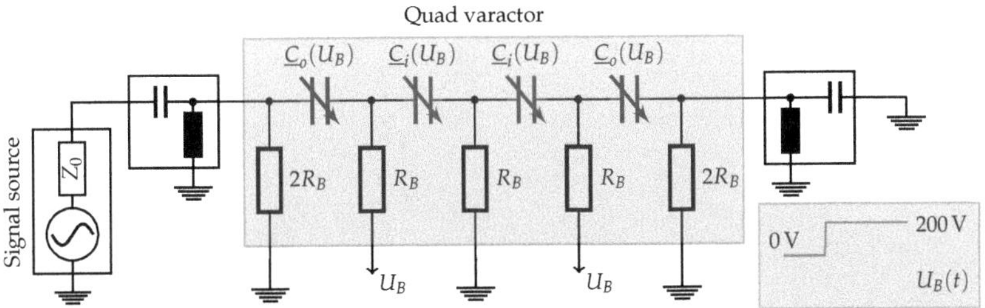

Figure 3.29: Equivalent circuit of a quad varactor.

$C_o = 0.5C_i$, results in:

$$P_o = \frac{C_i}{2C_i + 0.5(N-2)C_i} \cdot P_0 = \frac{1}{3}P_0 \ . \tag{3.39}$$

Under this assumption, the outer cell has to sustain 33 % of the total power (instead of 25 %).

Fig. 3.30(a) shows simulated voltage across the cells, which as a consequence leads to a different capacitive tuning shown in Fig. 3.30(b). The unbalance in capacitance, especially around $t = 1\,\mu s$ leads $C_o/C_i = 75\,\%$ and results in a capacitive voltage divider. According to (3.38), during that period of time, the outer cells have to sustain 28.5 % higher power level, compared to the inner cells. In total, both outer cells will have to sustain 57 % of the total power, which can lead to a destruction of the whole component, when operated close to maximum tolerated RF power.

The total fraction of RF power seen by the two outer varactors would reduce by further cascading N (3.38). However, the ratio of power it would have to sustain compared to a balanced case increases:

$$\frac{P_{\text{cell}}}{P_{\text{cell,balanced}}} = \frac{\frac{C_o(U_B)C_i(U_B)}{C_{\text{cell}}(U_B) \cdot (2C_i(U_B) + (N-2)C_o(U_B))} \cdot P_0}{\frac{P_0}{N}} \tag{3.40}$$

$$= \frac{C_o(U_B)C_i(U_B)N}{C_{\text{cell}}(U_B) \cdot (2C_i(U_B) + (N-2)C_o(U_B))} \ .$$

Under the same assumption that the outer capacitance has reduced to $0.5C_i$, (3.40) simplifies to:

$$\frac{P_o}{P_{o,\text{balanced}}} = \frac{N}{0.5N + 1} \ . \tag{3.41}$$

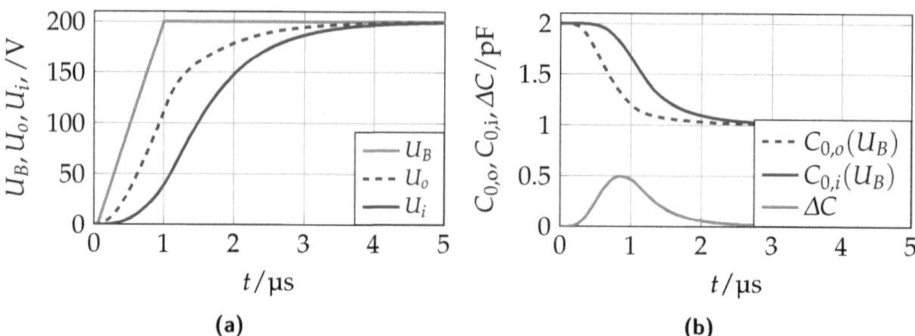

Figure 3.30: (a) Applied bias voltage $U_B(t)$ and the bias voltage across the outer $C_{0,o}(U_B)$ and inner $C_{0,i}(U_B)$ capacitor cells of the quad varactor. (b) The tuned capacitance $C_{0,o}(U_B)$ and $C_{0,i}(U_B)$ as a consequence of different bias voltage. The difference is given as ΔC.

In conclusion, the double varactors have only outer cells, which have the highest overall response and even power distribution. The quad varactor has the negative effect of additional power stress to the outer cells and has a reduced time response due to inner cells. Further cascading increases the power seen by each cell and, hence, increases the power handling capability of the varactor. However, the outer cells always have to face an increased power level, compared to a uniformly tuned varactor.

Inductive Decoupling

An alternative approach is using inductive decoupling by replacing resistors R_B with inductors L_b, yielding a complex capacitance

$$\underline{C}_{\text{u.c.}} = C_{\min} - \frac{1}{4\omega^2 L_0} - j\frac{C_{\min}}{Q_{\max}}. \tag{3.42}$$

In analogy to (3.34) the effective values are:

$$C_{\text{eff}} = \Re\left\{\underline{C}_{\text{u.c.}}\right\} = C_{\min} - \frac{1}{4\omega^2 L_b},$$

$$Q_{\text{eff}} = -\frac{\Re\left\{\underline{C}_{\text{u.c.}}\right\}}{\Im\left\{\underline{C}_{\text{u.c.}}\right\}} = Q_{\max}\left(1 - \frac{1}{4\omega^2 C_{\min} L_b}\right). \tag{3.43}$$

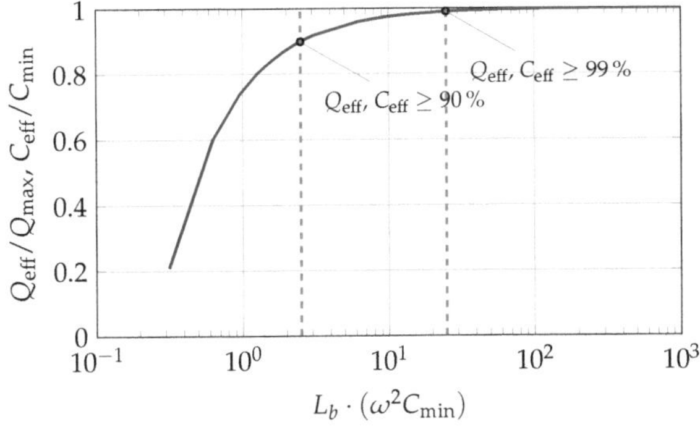

Figure 3.31: Impact of the bias inductance on the effective capacitance quality factor of a unit cell.

Fig. 3.31 shows normalized Q_{eff} and C_{eff} versus $L_b/(1/(\omega^2 C_{\text{min}}))$. To maintain 99 % of Q_{max} and C_{min} the bias inductance should be:

$$L_b \geq 25/\omega^2 C_{\text{min}}. \tag{3.44}$$

Assuming a capacitance with $C_{\text{max}} = 2\,\text{pF}$, $C_{\text{min}} = 1\,\text{pF}$, $Q_{\text{max}} = 100$ and $\omega = 2\pi \cdot 2\,\text{GHz}$ would result in $L_b = 160\,\text{nH}$. However, realization of an inductance with 160 nH having a resonance frequency SRF $> 2\,\text{GHz}$ is not trivial, therefore the threshold for L_b should be reduced to

$$L_b \geq 2.5/\omega^2 C_{\text{min}}, \tag{3.45}$$

which yields Q_{eff}, $C_{\text{eff}} \geq 90\,\%$ of the initial values with $L_b = 16\,\text{nH}$. When biasing with modulated signals one has to make sure that the resonance frequency of the RLC circuit is not excited, which would result in accumulation of energy and, hence, a bias voltage which might exceed the breakdown voltage of the capacitor.

Consequently, the minimum switching time or the modulation bandwidth of the bias signal should be calculated from the damping constant:

$$\alpha \geq \frac{1}{2R_0 C_{\text{min}}} + \underset{0}{\cancel{\frac{R_L}{2L_b}}} \equiv \frac{\omega}{2Q_{\text{max}}}. \tag{3.46}$$

Note that a lossless inductor is assumed and hence only the capacitive term remains in (3.46). This gives a minimum bias response time of $\tau = 16\,\text{ns} \equiv 62.5\,\text{MHz}$, which is significantly faster than the response time of resistive biasing. Effects described for resistive biasing when putting the varactors in real circuits are considered to be significantly less, since the decoupling inductors L_b are placed in parallel to similar components. However, in some circuits, asymmetric response can occur.

Depending on the targeted application and frequency, it can be beneficial to implement resistive biasing, as has been used for thin film varactors, see Section 3.2.2. Due to the series connection of 24 varactors, in combination with resistive biasing, the response time reduces down to 1 ms [STM15]. When targeting fast dynamic applications, the decoupling should be performed with inductors.

4 Large Signal Characteristics of Ferroelectric Varactors

This chapter treats the investigation of BST-based components under large signal operation. Two major effects occur when operating under high RF power:

1. RF-induced thermal drift due to dissipated power

2. RF-induced non-linearities

Both effects will be the focus of this section.

BST has a temperature dependent permittivity. Therefore, temperature increase in the substrate would lead to thermal capacitance drift, which should be avoided or reduced. Fig 4.1 shows the characterization results of a Cu-F doped BST thick film layer over temperature T at 5 GHz, obtained from a CPW characterization method described by [Sch07a].

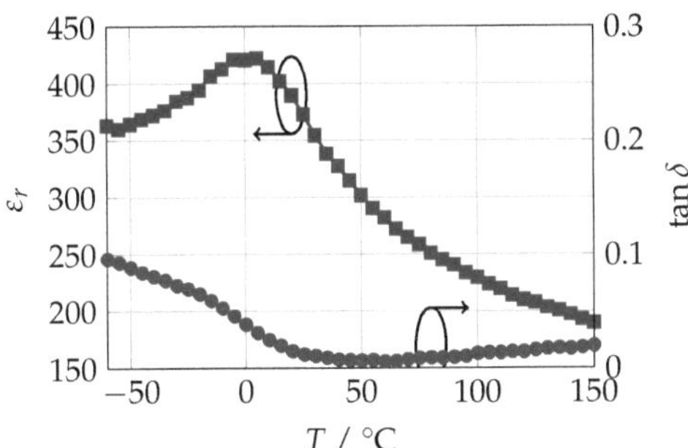

Figure 4.1: Dielectric properties of a Cu-F codoped BST layer over a temperature range from −60 °C to 150 °C, extracted at 5 GHz.

4.1 Thermal Conductivity of Barium Strontium Titanate

In practical applications, high Q components are desired but are not always available, especially when working with ferroelectrics. Under high RF power operation, the components transfer electromagnetic energy into thermal heat proportional to $\tan\delta = R_s/(\omega C)$ where R_s is the equivalent series resistance. Under RF power operation, a temperature increase of the component is expected until an equilibrium between the dissipated energy and the removed thermal energy is reached. The dissipated heat is transferred from the capacitor by means of:

1. thermal radiation

2. transfer of heat through air (convection)

3. transfer of heat from the substrate to a heat sink

The latter contributes most and, therefore, the thermal conductivity κ of the substrates is of utmost importance to allow statements on temperature increase and high power tolerance of the ferroelectric capacitors. A widely used method for thermal characterization is the light flash method, which has been proposed by [Par+61]. Fig. 4.2(a) shows the measurement setup with a sample placed in an isolated container. To allow the characterization under various temperatures, the container can be heated or cooled. The initial sample temperature is monitored with a sample thermocouple. The sample is exposed to a high power laser pulse, which heats the backside of the sample. The top side temperature is monitored with a detector. Fig. 4.2(b) shows a typical signal obtained over time.

From the saturation point $\Delta T_{\max}$ the specific thermal capacity c_p of the material can be extracted with the relation:

$$\Delta T_{\max} \propto \frac{1}{c_p} \ .$$

(4.1)

Further, the thermal diffusivity α can be extracted with the relation:

$$\alpha = \frac{1.38d^2}{\pi^2 t_{1/2}} \ ,$$

(4.2)

where d is the thickness of the sample and $t_{1/2}$ is the time required for the top surface to reach half of the maximum temperature rise $\Delta T_{\max}$. Knowing the thermal

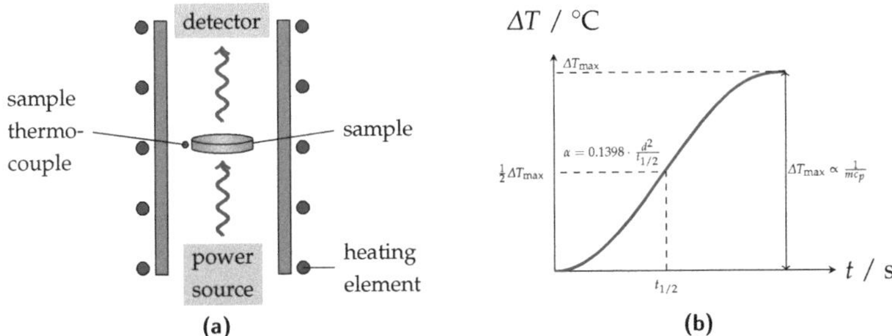

Figure 4.2: (a) Schematic of the Light Flash-Method (b) Exemplary detector curve to extract c_p and α of the sample.

diffusivity α, the specific thermal capacity c_p and the density ρ of the sample, the thermal conductivity κ can be calculated as:

$$\kappa(T) = \alpha(T) \cdot \rho(T) \cdot c_p(T) \ . \tag{4.3}$$

This method was used to analyze four materials over a temperature range from 50 °C to 500 °C with the LFA 457 MicroFlash® device. Aluminum nitride (AlN) and alumina (Al_2O_3) are investigated as carrier substrates. Iron fluorine (Fe-F) doped $Ba_{0.6}St_{0.4}TiO_3$ samples were fabricated as bulk and as a 4 µm screen-printed layer on top of an alumina carrier substrate. The latter materials are porous and to determine the materials density, the buoyant force is employed by putting the samples in water and measuring their weight. Fig. 4.3 summarizes the thermal conductivity κ for each material, where two samples of each have been measured. The thermal conductivity of AlN is one order of magnitude higher than pure alumina, which is consistent with the values reported in literature [KUT88]. The bulk BST sample shows a thermal conductivity in the range of 10 W/(m K), which is factor three less than alumina. The substrates coated with a BST layer show a reduction of approximately factor 2, which implies a significantly worse thermal conductivity of the screen-printed BST layer compared to the bulk sample. This can be explained by the higher porosity of the screen printed material. To estimate the thermal conductivity of this layer a model is assumed, where each layer contributes to the total thermal conductivity κ_{eff} as [AKM88]:

$$\frac{\sum_i d_i}{\kappa_{eff}} = \sum \frac{d_i}{\kappa_i} \ , \tag{4.4}$$

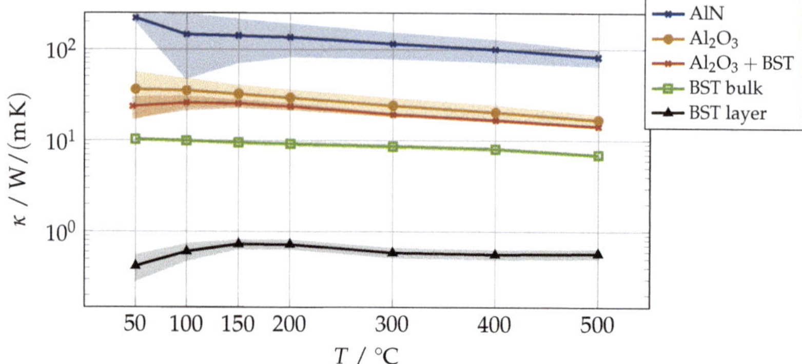

Figure 4.3: Thermal conductivity κ of AlN, Al_2O_3, bulk BST and Al_2O_3 substrate coated with a 4 µm thick BST layer. The conductivity of the BST layer was extracted from an equivalent resistance model. The shaded areas denote the uncertainty of the measurements and calculations.

where d_i is the thickness of each layer respectively. In this equation, the components are treated as layers of same size running parallel to the surface, providing an insulating effect that is dominated by the component with lowest conductivity. Knowing κ_{eff}, $\kappa_{alumina}$ of pure alumina and the thickness of both layers (4.4) yields:

$$\kappa_{BST} = \frac{4\,\mu m}{\frac{639\,\mu m}{\kappa_{eff}} - \frac{635\,\mu m}{\kappa_{alu}}} \, . \tag{4.5}$$

Fig. 4.3 shows the resulting thermal conductivity of the BST layer around $0.5\,W/(m\,K)$. These values will be used for the following high power investigations.

4.2 RF-induced Self Heating

The RF-induced self-heating of the IDC varactors has been previously investigated by [Mau11] with the result that an IDC varactor should sustain 50 W RF power while heating up by approximately $20\,°C$. However, the thermal conductivity of the BST layer was assumed similar to alumina, and is order of magnitude higher than the values found in Section 4.1. Further, it is not clear whether the 50 W RF power is the accepted power or the available source power. Therefore, a multi-physics

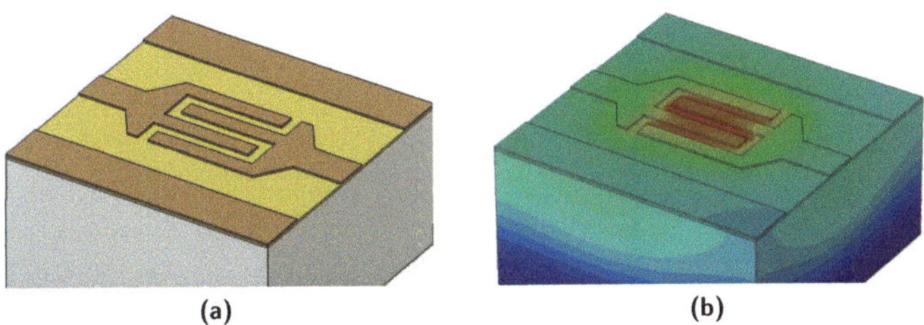

Figure 4.4: (a) The IDC model implemented in CST Studio Suite (b) Results of the thermal co-simulation on alumina. The temperature between the IDC fingers has risen by approximately 10 °C.

simulation in CST Studio Suite was implemented with the typical layer structure, shown in Fig. 3.23, and the thermal conductivity values for the substrates layers, found in the previous section.

4.2.1 Modelling of Thermal Behaviour

A two-port EM time domain simulation is performed to obtain field and current distribution of the component under investigation. For the simulation, an IDC with a finger length $l = 160\,\mu\text{m}$, $w = 30\,\mu\text{m}$, $g = 20\,\mu\text{m}$ and a number $N = 4$ of fingers was used. The conducting layer was set to $h = 3\,\mu\text{m}$ with the conductivity $\sigma = 4.1 \cdot 10^7\,\text{S/m}$. The BST layer dielectric properties were set to $\varepsilon_r = 400$ and $Q = 50$. The simulated series varactor, shown in Fig. 4.4(a), creates an impedance mismatch between both ports. This effect has to be considered to calculate the power the component really is exposed to.

The simulation feeds the ports with a total RF power of 0.5 W, while the actual input RF power is in the range of 0.01 W. The remaining power is reflected back into the source. Therefore, a power scaling factor $S = 50$ was used to make sure that the device is exposed to approximately 0.5 W. The results are shown in Fig. 4.4(b). A temperature increase in the BST layer between the IDC fingers of approximately 10 °C is predicted by the simulation. This value is approximately factor 100 higher than the one obtained by previous investigation [Mau11].

Further, a carrier substrate analysis was performed by changing the thermal conductivity to the value of AlN, which is one order of magnitude higher than alumina.

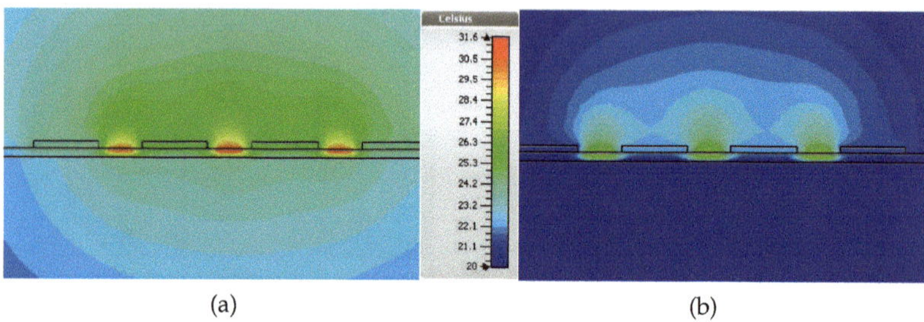

Figure 4.5: (a) Temperature profile of the IDC on alumina carrier substrate. (b) Temperature profile of the IDC on AlN carrier substrate. The scale of both simulations is equal and placed between both pictures.

The comparison between alumina and AlN is summarized in Fig. 4.5. Overall, a significant temperature reduction can be observed. However, especially the areas, where the heat is generated, do not show any significant temperature reduction. This effect can be attributed to the relatively low thermal conductivity of the BST layer. It can be concluded, that no thermal drift reduction can be achieved by the use of highly thermal conductive carrier substrates as long as the BST layer is the limiting factor.

The same investigations were carried out for a MIM topology, which are summarized in Fig. 4.6. This topology is expected to sustain higher RF power since the bottom electrode has direct contact to a highly thermal conductive substrate and can transfer the heat directly into the carrier substrate.

However, the temperature increase predicted by the simulation is higher than for the IDC case, which is not surprising since the total dissipating volume is significantly smaller. With a total accepted RF power of 0.5 W a temperature increase of $20\,°C$ on alumina and $15\,°C$ on AlN carrier substrate are predicted in a volume of $4\,\mu m \times 50\,\mu m \times 50\,\mu m$. Considering a maximum tolerable temperature increase of $10\,°C$ and a constant BST layer thickness of $4\,\mu m$ a rated RF power handling capability of

$$P_{\text{rated}} = 100\,\text{W/mm}^2, \quad \text{Alumina}$$
$$P_{\text{rated}} = 130\,\text{W/mm}^2, \quad \text{AlN}$$

(4.6)

can be calculated for the MIM topology. However, with a permittivity of $\varepsilon_r = 400$ and a BST layer thickness $h = 4\,\mu m$, the resulting capacitance would be too large when designing varactors for high power applications. Therefore, either thicker BST

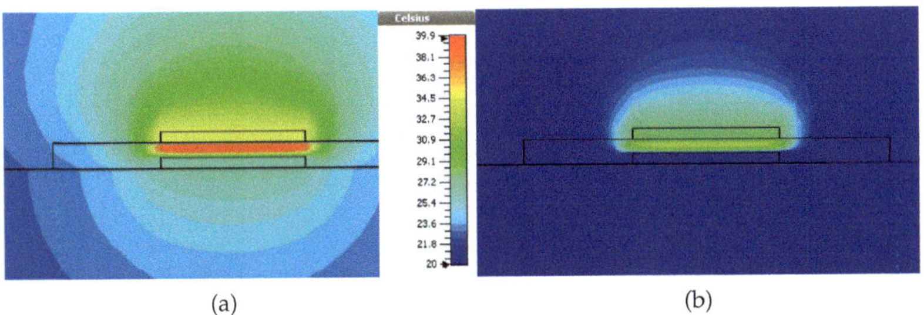

(a)   (b)

Figure 4.6: (a) Temperature profile of the MIM varactor on alumina carrier substrate (b) Temperature profile of the MIM varactor on AlN carrier substrate.

layers or a series connection of MIM varactors is inevitable to fabricate varactors suitable for the GHz frequency range. To calculate the required number of series unit cells N for a total capacitance C_{tot} one can use:

$$C_{tot} = C_0/N = \varepsilon_0 \varepsilon_r \frac{A_0}{d}/N \ , \tag{4.7}$$

with a single cell capacitance C_0, where A_0 is the capacitor area of the cell and d the BST layer thickness. The rated power handling capability yields a required unit cell area of:

$$A_0 = \frac{P_{req}}{P_{rated}}/N \ , \tag{4.8}$$

where P_{req} is the required power handling capability of the varactor. Combining (4.7) and (4.8) leads to a required number of varactor cells of:

$$N = \sqrt{\varepsilon_0 \varepsilon_r \frac{P_{req}/P_{rated}}{C_{tot}d}} \ , \tag{4.9}$$

with $N \in \mathbb{N}$.

Knowing N, the required unit cell area can be directly calculated from (4.8). A tunable varactor with $C_{tot} = 2\,\text{pF}$ and a required power handling capability of e.g. $10\,\text{W}$ would require $N = 6.6 \rightarrow N = 7$ series varactors with a unit cell area $A_0 = 125 \times 125\,\mu\text{m}^2$. However, the low temperature sintered BST usually has a lower permittivity in the range of 100–200. Therefore, lower number of varactors would be required in practice.

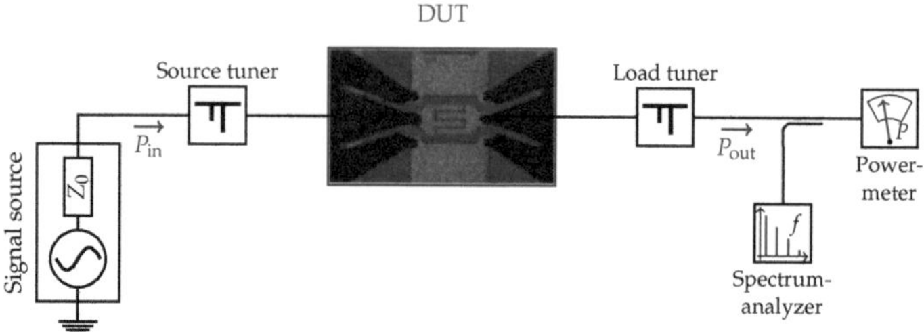

Figure 4.7: Large signal measurement setup for the tunable varactor. 150 µm GSG-probe are used to contact the varactor.

4.2.2 Large Signal Measurements

The large signal behavior of the BST varactors was investigated using large signal load-pull (LP) measurement setup similar to those used for the investigation of high power transistors. Fig. 4.7 shows the schematic of the setup. The device under test (DUT) is placed between two mechanical impedance tuners to ensure matched condition between the signal generator and the power meter. Two ground-signal-ground (GSG) probes with 150 µm pitch are used to contact the varactor. A spectrum analyzer is used to keep track of the linearity of the system.

IDC

A single series IDC varactors on 635 µm thick alumina substrate coated with a 4 µm thick Cu-F doped BST layer is investigated. The varactor was kept unbiased at $U_B = 0\,\text{V}$ and mechanically tuned to match the $50\,\Omega$ of the source impedance. The load was matched for maximum output power at $T = 20\,°\text{C}$ with an available source power (P_{avs}) of 10 dBm, or small signal conditions at the input. These impedance conditions were kept constant and continuous wave (CW) power sweeps were conducted. The result is shown in Fig. 4.8. The transducer power gain G_T was calculated as the difference between the measured output power P_{out} and the measured input power P_{in} after correcting the contributions from cables, couplers, mechanical tuners and probes.

$$G_T = P_{\mathrm{out}} - P_{\mathrm{in}} + \sum \mathrm{IL}_{\mathrm{setup}} [\mathrm{dB}] \ . \tag{4.10}$$

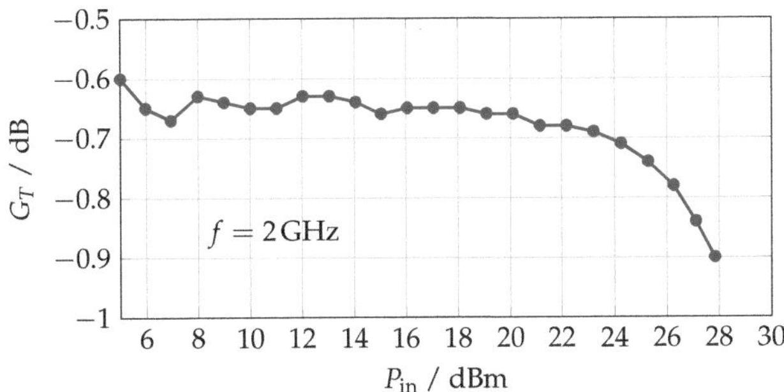

Figure 4.8: Large signal power sweeps of a single series IDC varactor at $T = 20\,°\mathrm{C}$.

Figure 4.9: (a) RF-induced thermal destruction of an IDC varactor and a MIM varacotr. (b) The temperature was sufficient to melt the gold electrodes.

With increasing power level, the transducer power gain $G_T = -0.65\,\mathrm{dB}$[1] is gradually reducing, see Fig. 4.8. Especially around 23 dBm a strong reduction of G_T is observed. This effect can be attributed to an impedance shift, which is the consequence of a thermal permittivity change of the BST layer and, hence, the capacitance which leads to an impedance mismatch between the source and the power meter. The mechanical tuners can be used to correct this mismatch, which was not done here, since the reduction of the transducer gain is used to draw conclusions on the temperature increase of the DUT. However, adjusting the impedance mismatch and further increasing the RF power, ultimately, leads to thermal destruction of the device, as shown in Fig. 4.9(a).

---

[1]Note that in small signal limit the varactor is perfectly matched. Therefore the insertion loss of the varactor can be readily read as $\mathrm{IL} = 0.65\,\mathrm{dB}$.

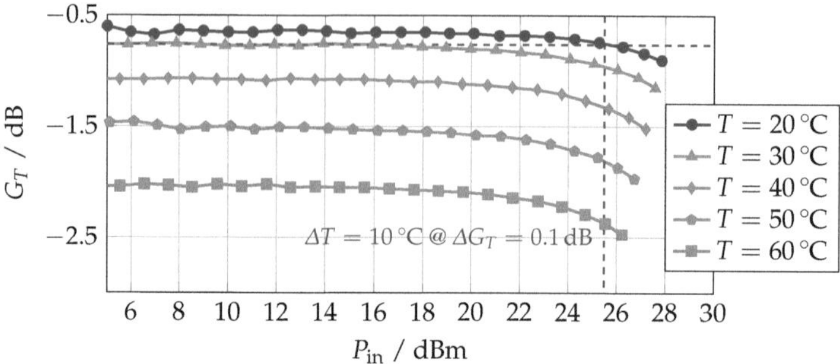

Figure 4.10: Large signal power sweeps at 2 GHz of a single series IDC varactor at different temperatures while keeping matched condition for the value at $T = 20\,°C$.

To draw conclusions on the temperature increase from the dissipated power in the BST layer, power sweeps were conducted at different temperatures, while keeping the matched condition obtained at $T = 20\,°C$. Fig. 4.10 shows the extracted results.

Assuming that the reduction of G_T is mainly caused by an impedance mismatch, as a consequence of a temperature increase of $\Delta T = 10\,°C$ with $\Delta G_T \approx 0.1\,\mathrm{dB}$, the dissipated RF power P_D can be calculated as:

$$
\begin{aligned}
P_D &= P_\mathrm{in} - P_\mathrm{out} \\
&= \left(10^{(25.5\,\mathrm{dBm}+0.65\,\mathrm{dB})/10} - 10^{(25.5\,\mathrm{dBm})/10} \right)\,\mathrm{mW} = 57\,\mathrm{mW}\ .
\end{aligned}
\tag{4.11}
$$

It can be concluded that a temperature increase of $\Delta T = 10\,°C$ is caused by a dissipated RF power of 57 mW, which is in good agreement with the results obtained from CST simulation.

To fully prove that this mechanism is caused by RF induced self-heating of the varactor and not due to a field induced self-biasing, the measurements at $T = 60\,°C$ were repeated under pulsed RF conditions. A pulse width of 1 µs was used for near isothermal conditions and the duty-cycle (D) was varied from 10 % to 50 % to 100 %. The results are shown in Fig. 4.11.

With a reduced duty-cycle, the dissipated power is reduced, which as a consequence clearly reduces the power-induced G_T shift, affirming initial assumption of a dominating thermally induced impedance shift of the varactor. The turn-on

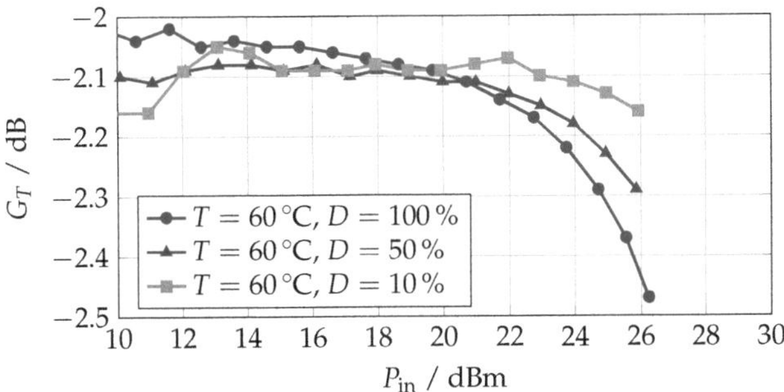

Figure 4.11: A set of power sweeps at 2 GHz conducted with a pulsed RF source and a pulse width of 1 µs at various duty-cycles D.

of G_T reduction is not avoided fully according to the duty-cycle 3 dB (50 %) and 10 dB (10 %), respectively. That can be attributed to the 1 µs pulse-width, which is not sufficiently short for fully isothermal conditions. During the 1 µs pulse, some self-heating is present and is sufficient to slightly heat and, therefore, slightly shift the impedance of the varactor.

MIM

The IDC in Fig. 4.7 was replaced by MIM varactors shown in Fig. 4.12. The BST layer was measured with a relative permittivity of approximately 200 and a layer thickness around $h = 8\,\mu m$. Three different topologies were used to investigate the power handling capability. For a comparison with simulations, a single MIM varactor cell, shown in Fig. 4.12(a), with a capacitance $C_s = 0.46\,pF$ and an area of $A_0 = 50 \times 50\,\mu m^2$ was used. The scaling of the power handling capability mentioned in Section 4.2 has been investigated with a double-cell and a quad-cell varactor, which are depicted in Fig. 4.12(b) and Fig. 4.12(c), respectively. However, it is important to note, that the size and shape of the capacitance intrinsically determines the setting of the load tuner for matched condition. This, on the other hand, determines the voltage across the varactor and hence the dissipated RF power. Therefore, the scaling of the varactors was performed such, that the capacitance of the single-, double- and quad-cell varactors are similar and the load tuner is kept at a constant impedance. Fig. 4.13 summarizes the power sweeps across the varactors shown in Fig. 4.12.

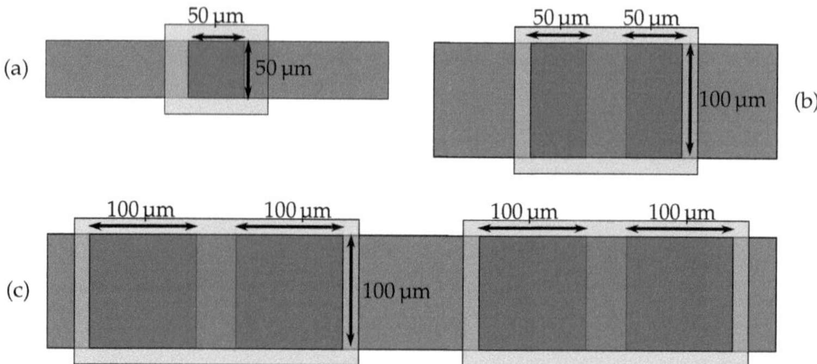

Figure 4.12: (a) Single cell varactor with $50\,\mu m \times 50\,\mu m$ overlap region with capacitance $C_s = 0.46\,pF$. (b) Double-cell varactor with $2 \times 100\,\mu m \times 50\,\mu m$ overlap region and a capacitance $C_d = 0.55\,pF$. (c) Quad-cell varactor with $4 \times 100\,\mu m \times 100\,\mu m$ overlap regions and a total capacitance $C_q = 0.57\,pF$.

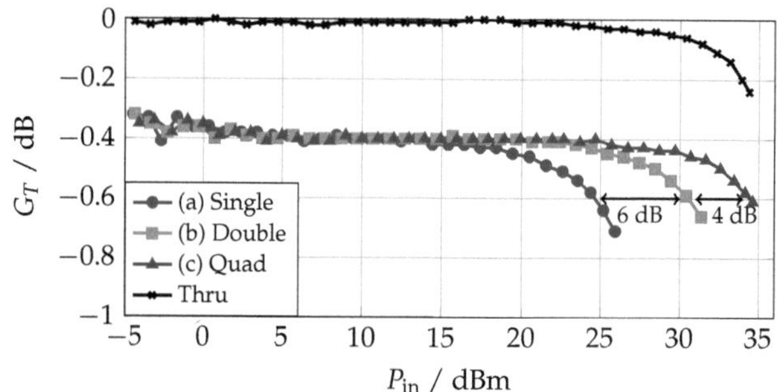

Figure 4.13: Transducer gain G_T of the MIM varactors at $2\,GHz$. The legend refers to the varactors in Fig. 4.12. The thru measurement reveals a bias tee related thermal drift, which is later used to correct the actual measurements.

An increase of power handling capability of approximately $6\,dB$ between the single-cell and the double-cell varactor is observed. An improvement of $3\,dB$ can be attributed to the capacitive voltage division between the series varactors. The additional $3\,dB$ are due to the increase of the capacitive area, which reduces the power

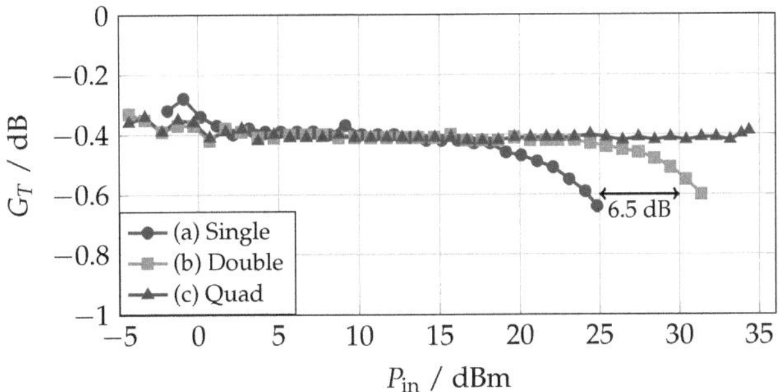

Figure 4.14: Corrected measurements of Fig. 4.13 at 2 GHz. The legend refers to the varactors in Fig. 4.12.

density by a factor of two. However, the double- and the quad-cell varactors did not show the expected scaling of power handling capability of 6 dB. A measurement of a thru revealed the limitations of the measurement setup, see Fig. 4.13. Above 30 dBm, a clear measurement setup related thermal drift is observed, which is caused by the utilized bias tees. The data was corrected and is plotted in Fig. 4.14. The power handling capability of the quad-cell varactor exceeds the available source power and can be assumed to follow the 6 dB increase. The deviation from the exact 6 dB scaling is expected to be caused by slightly varying capacitance values in combination with varying line widths and lengths.

Further, the power was re-normalized to W/mm^2, see Fig. 4.15. A power handling capability of MIM varactors $P_{rated} > 50\,$W/mm^2 can be expected, which is in good agreement with the thermal simulation (4.6) and is at least one order of magnitude higher than for the IDC varactors. In accordance to the IDC varactors, increase of RF power, ultimately leads to a thermal destruction of the component as shown in Fig 4.9(b).

Thin film Varactors

Since 2014 STMicroelectronics markets thin film varactors with the specifications given in [STM15]. The claimed properties have been verified by in-house measurements, see Appendix B. By serially cascading up to 24 varactors (Fig. 3.24), the

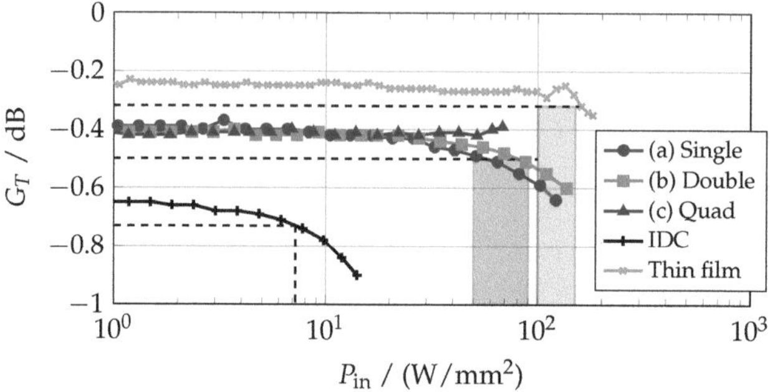

Figure 4.15: Normalized power density of IDC, MIM and commercial thin film varactors at 2 GHz .

linearity and high power handling capability of thin film varactors has been improved. However, the manufacturer implements resistive biasing concept described in Section 3.2.3. Therefore, the improvement of linearity and power handling is gained at the expense of tuning speed, which is in the range of $3\tau \approx 1\,\text{ms}$. The high power handling capability of the thin film varactors has been measured in the measurement setup shown in Fig. 4.7 and is put into context in Fig. 4.15. A $P_{\text{rated}} = \left[100\,\text{W/mm}^2 \text{ to } 150\,\text{W/mm}^2\right]$ has been estimated for the commercial varactors, which is up to a factor 2 higher compared to the thick film MIM varactors. However, a final conclusion cannot be drawn as:

1. The measured capacitance value differs from the MIM, which requires a different load tuner setting for matched condition. Therefore, the voltage drop across the varactors changes and, hence, the dissipated RF power.

2. The commercial varactors are processed not only planar but also utilize the third dimension. For the calculation of P_{rated} only the lateral dimensions were estimated with $A_0 = 8\,\mu\text{m} \times 195\,\mu\text{m}$. However, the exact area could not be measured.

The onset in G_T of 0.2 dB implies a higher Q of thin film varactors due to purer and more homogenous BST layer, which has been supported by small signal measurements. Around 2 GHz thin film varactors show $Q \approx 45$, while the thick film MIM varactors show typically $Q \approx 20$. The small signal characterization results can be found in the Appendix B.

4.3 Non-linear Characteristics

Current, voltage or field dependent components intrinsically suffer from non-linear effects. Assuming a sinusoidal two-tone signal

$$x(t) = \Re \left\{ A_0(e^{j\omega_a t} + e^{j\omega_b t}) \right\} \; , \tag{4.12}$$

a non-linear device introduces a non-linear transformation $y(t) = H(x(t))$. To approximate this non-linear transformation function, the Taylor series can be used. Consequently, the response of the input becomes

$$y(t) = k_0 + k_1 \cdot x(t) + k_2 \cdot x(t)^2 + \cdots + k_n \cdot x(t)^n \; , \tag{4.13}$$

with n being the order of the intermodulation product (IMn). However, the even orders intermodulation products can be easily filtered, while the odd order intermodulation products contain portion of frequencies, which can directly fall into the operation band and cannot be easily filtered, as they lie closely to the band of operation. The amplitude of the IM products also strongly depends on the order and is typically strongly decreasing with increasing order. Hence, third order intermodulation product (IM3) can be considered as the most critical and is often used to benchmark the linearity of devices.

4.3.1 Models

To model the non-linear behavior of voltage-tunable BST varactors, harmonic balance simulation in ADS is used. Unlike varactor diodes, the characteristic capacitance-voltage curve of BST is symmetric and therefore has to be tailored only with even coefficients.

$$C(U_B) = \sum_{n=0}^{N} c_{2n} U_B^{2n} \; . \tag{4.14}$$

The characteristic curve can be obtained from small signal measurements, fitted with (4.14) to obtain the coefficients c_{2n} and can be implemented into a non-linear capacitance model of ADS. Fig. 4.16a exemplary shows a small signal characteristic curve obtained with (2.17) fitted with (4.14) with the even order $N = 9$. Note, that for large voltages U_B, the fit becomes inaccurate, which is caused by numerical inaccuracies in combination with great powers $2n$. Therefore, to accurately model the non-linear behavior in a harmonic balance simulation, the RF voltage swing should be well below these values (in this example $< 150\,\mathrm{V}$).

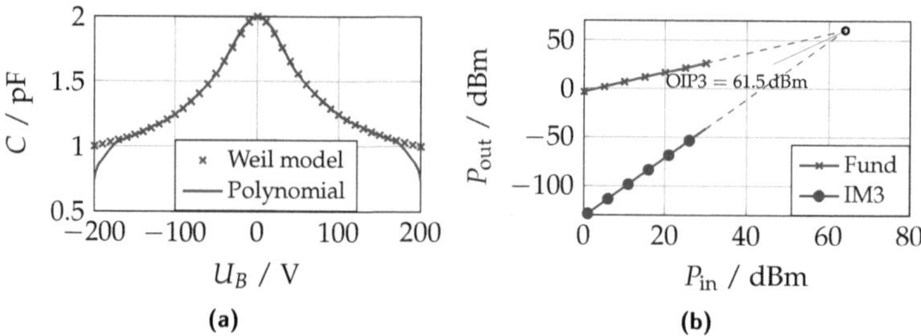

Figure 4.16: (a) Exemplary characteristic capacitance curve obtained with the model proposed by Weil and fitted with even polynomials up to $N = 9$. (b) Fundamental and IM3 power levels obtained with harmonic balance simulation.

In typical applications, the intermodulation power is plotted versus the total input power. However, in case of a voltage-tunable capacitor, the RF voltage is the source of the non-linearity. Therefore, the implementation of the tunable capacitor plays an important role. The varactor can be placed in a resonant structure, which accumulates energy, and hence the voltage across the capacitor. Consequently, large intermodulation products even for small RF power levels can occur. In general three cases have to be considered:

1. overdamped case with $1/(\omega C) < R$,

2. critically damped case $1/(\omega C) \approx R$,

3. underdamped case with with $1/(\omega C) > R$.

For the work here, only non-resonant circuits are of interest, therefore, only the overdamped case is considered.

Fig. 4.16b shows simulated fundamental and third order intermodulation product (OIP3) obtained with the harmonic balance simulation for tunable capacitance characteristic shown in Fig. 4.16a. A two-tone signal with 1 MHz tone spacing around 2 GHz was used and the total RF power was swept from 0 dBm to 30 dBm. An OIP3 of 60.5 dBm was obtained by extrapolation at $U_B = 0$ V.

Scaling of Linearity

Changing the capacitance generally does not change the characteristic shape of the C/V curve but can be considered as a mere scaling by a factor $S = C/C_0$. Consequently, the voltage drop across the varactor changes. Considering an overdamped RLC circuit, the voltage across the capacitor U_c scales anti proportional to the capacitor value:

$$U_c \propto \frac{1}{C}. \tag{4.15}$$

Thus, to achieve the same IM3, the RF power has to change by

$$\Delta P_{\text{in}} = 20 \log \left(\frac{C}{C_0} \right). \tag{4.16}$$

Knowing that the IM3 is increasing with a slope steepness of 3, leads to a change of OIP3 according to:

$$\begin{aligned} \Delta OIP3 &= 1.5 \cdot \Delta P_{\text{in}} \\ &= 30 \log \left(\frac{C}{C_0} \right) \\ &= 30 \log \left(S \right). \end{aligned} \tag{4.17}$$

Scaling the capacitance $C_0(U)$ to any capacitance value $C(U) = S \cdot C_0(U)$, changes the known $OIP3_{C_0}$ according to:

$$\begin{aligned} OIP3 &= OIP3_{C_0} + \Delta OIP3 \\ &= OIP3_{C_0} + 30 \log \left(S \right). \end{aligned} \tag{4.18}$$

Similar considerations can be carried out for the case of N concatenated varactors with the single cell capacitance C_N and a total capacitance C_{tot}/N. The chain of varactors functions as a capacitive voltage divider. Consequently, the voltage across a single varactor reduces to $U_c = \frac{U_0}{N}$. Comparing a single call varactor with a nominal capacitance C_0 and a series of varactors with the same capacitance, the OIP3 of each individual capacitance cells scales as:

$$OIP3_N = OIP3_{C_0} + 30 \log N. \tag{4.19}$$

However, each of the nonlinear varactors in the chain produces the intermodulation. Assuming that the individual intermodulation occurs in-phase, the total OIP3 of the series-cell varactor, compared to a single-cell implementation, scales with:

$$\begin{aligned} OIP3_{\text{tot}} &= OIP3_{C_0} + 30 \log N - 10 \log N \\ &= OIP3_{C_0} + 20 \log N. \end{aligned} \tag{4.20}$$

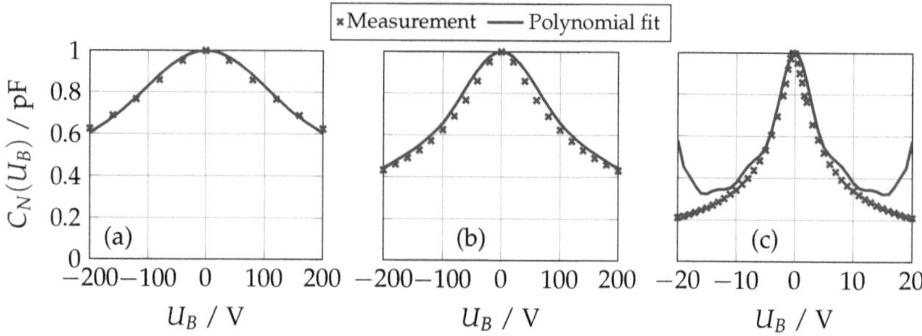

Figure 4.17: Normalized capacitance curve $C_N(U_B)$ of different varactor implementations – (a) IDC 1 pF, (b) MIM 2 pF and (c) thin film 5 pF, extracted from small signal measurements. Polynomial fits of order $N = 9$ have been applied.

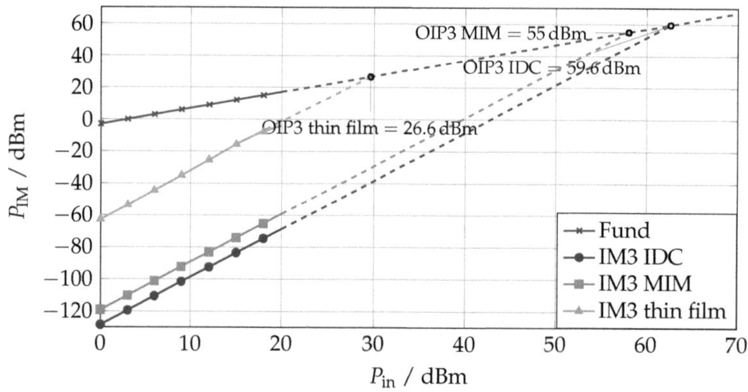

Figure 4.18: Simulated fundamental and intermodulation product power of the IDC, MIM and STMicroelectronic thin film varactor at $U_B = 0\,\text{V}$. Polynomial fit parameters of the C/V curves from Fig. 4.17 were used.

To model the non-linear effects of different varactor implementations, a small signal C-V characteristic is required. Fig. 4.17 shows measured data of an IDC, MIM and thin film varactor normalized to $C = 1\,\text{pF}$ and fitted with (4.14) up to $N = 9$.

Each varactor model polynomial was used in a harmonic balance simulation to extract the OIP3, see Fig. 4.18.

The thick film technology (IDC and MIM) show the highest OIP3. Here, the IDC varactors feature higher linearity compared to the MIM implementation due to higher bias voltages, which is the consequence of large IDC finger gap ($g \approx 15\,\mu m$), compared to the BST layer thickness ($d \approx 8\,\mu m$) of the MIMs. The thin film implementation shows a significantly lower linearity, which appears unpractical. However, the thin film varactor features a total of $N = 24$ series cells, therefore, the value obtained in the simulation has to be scaled in accordance to (4.20), leading to an OIP3 of 54.2 dBm, which is in the expected range.

4.3.2 Intermodulation Measurements and Comparison

The linearity of the varactors was investigated in a CW two-tone measurement at 2 GHz with 1 MHz tone-spacing (1.9995 GHz, 2.0005 GHz), using the large signal measurement setup shown in Fig. 4.7. The varactors were driven in small signal matched condition at room temperature in the unbiased state. Fig. 4.19 shows the measured fundamental power and the IM3 for the three different varactor realizations – IDC, MIM and thin film.

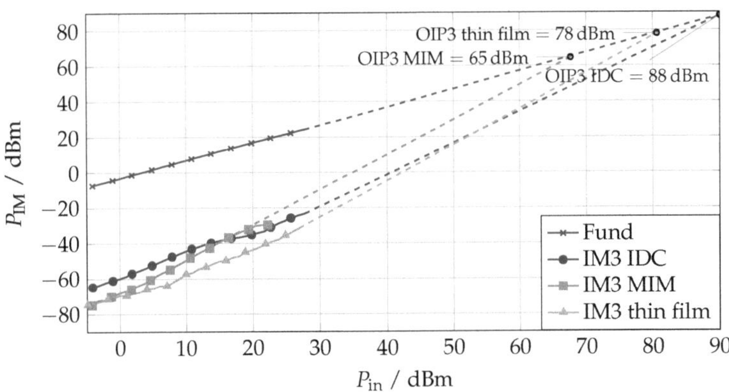

Figure 4.19: Measured fundamental and intermodulation product power of the IDC, MIM and STMicroelectronic thin film varactor.

Table 4.1: Measured capacitance C, OIP3 and the normalized OIP3_N of the different varactor implementations as well as the normalized simulated $\text{OIP3}_{N,\text{sim}}$.

	C [pF]	S	OIP3 [dBm]	OIP3_N [dBm]	$\text{OIP3}_{N,\text{sim}}$ [dBm]
IDC	1	1	88	88	60
MIM	0.5	2	65	74	55
Thin film	5	0.2	78	64	54

The measurement does not show the expected slope of 3 for the IM3 but varies between 1.7 for IDC and 2.2 for thin film varactors. Currently, it is believed that this effect is caused by a combination of insufficient linearization of the measurement setup, and the high linearity of the devices under test. Even for power levels above 20 dBm the contribution of the varactors appears to be negligible. Taking this into consideration, it can be concluded that the simulated OIP3 values should give the lower limit.

However, the varactors used in the measurement have different capacitance value. Therefore, to compare the results, the obtained OIP3 was normalized in accordance to (4.18). Table 4.1 summarizes the results. The measurement, though not fully evident due to limitations of the measurement setup, shows the same trend as the simulated results. The IDC varactors exhibit the highest linearity, followed by MIM and the cascaded thin film varactor implementation. It can be concluded that the values are high enough and have negligible impact on the linearity of a power transistor, which has a typical OIP3 of approximately 40 dBm.

5 Integration of Tunable Matching Networks into RF Power Amplifiers

The transfer of power from a generator to a load constitutes one of the fundamental problems in the design of communication systems. A problem of this type involves in every case the design of a (lossless) coupling network to transform an arbitrary, complex load impedance $\underline{Z}_L$ into another specified impedance, typically 50 Ω. One refers to this operation as *impedance matching*. In most practical cases, this problem can be idealized with lossless networks, as indicated in Fig. 5.1. Such networks usually constitute of fixed lumped elements. The quality of matching of circuits, utilizing fixed elements, is theoretically limited, which has been first investigated by Bode[Bod55] and later generalized by Fano[Fan48]. It has been shown, that the quality of matching and the bandwidth are interdependent, which limits the matching quality over broad frequency span.

This section covers the investigation and limitation of *tunable* impedance matching networks (TMN) based on ferroelectric varactors to overcome these theoretical limitations. The matching is carried out only in a fractional bandwidth, which can be tuned. This effectively allows access to broader frequency spectrum, without reducing the quality of the matching within the fractional bandwidth. Further, the implementation of the TMNs into high power RF transistors is investigated. When placed at the output of a transistor, tunable networks potentially allow to improve the efficiency or maximum available output power over frequency (Bode-Fano) or input power dynamics by readjusting the matching.

Figure 5.2a shows the normalized power distribution of an OFDM modulated signal with 32 sub-carriers, combined with a simulated power added efficiency (PAE) of a gallium nitride (GaN) high electron mobility transistor (HEMT), driven in class AB at 2 GHz. The power added efficiency is defined as:

$$\text{PAE} = \frac{P_{\text{s,out}} - P_{\text{s,in}}}{P_{\text{DC}}} \ . \tag{5.1}$$

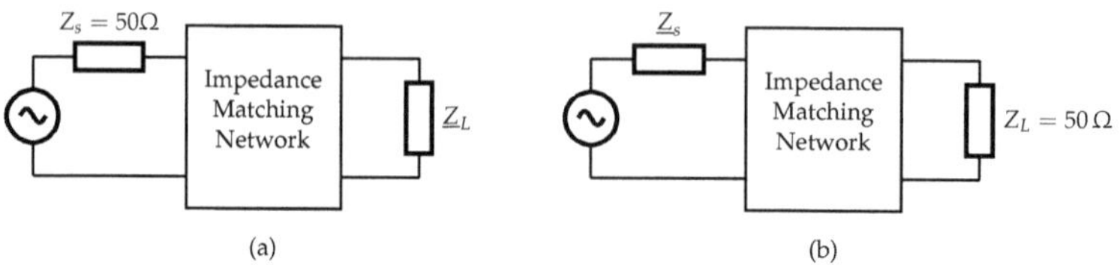

Figure 5.1: Idealized matching of a fix source impedance to an arbitrary complex load (a) and vice versa (b).

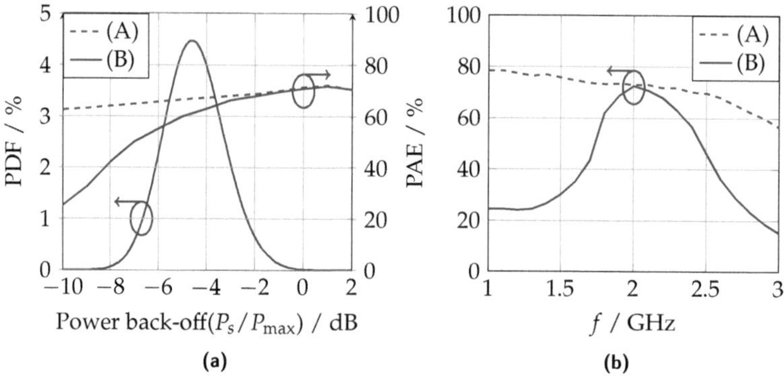

Figure 5.2: (a) Normalized power distribution of an OFDM signal with 32 sub-carriers, combined with simulated PAE of a GaN HEMT with varying (A) and fixed load (B). (b) PAE distribution versus operation frequency of a GaN HEMT with varying (A) and fixed load (B).

In this figure, two scenarios have been assumed: (A) the load for the transistor is optimized for each power level individually and (B) the load impedance is optimized for 0 dB power back-off and kept constant, while increasing the power back-off level.

The dynamic power added efficiency $\eta_{\text{eff,P}}$ can be calculated as [Mau11]:

$$\eta_{\text{eff,P}} = \int\limits_{P_s=0}^{P_s=\infty} \text{PDF}(P_s) \cdot \text{PAE}(P_s) \, dP_s \ . \tag{5.2}$$

In this simplified example, the tuned case results in an overall efficiency improvement of $\Delta\eta_{\text{eff,P}} = 6\,\%$.

Beside the TMN approach, which is the focus of this work, several other methods have been proposed and investigated to target this circumstance. Supply modulation, employing envelope tracking, is used to increase the dynamic efficiency [Kah52; Kim+08] without using output impedance variation. Alternatively, the Doherty active load modulation approach is widely used to increase the power dynamics efficiency at the expense of bandwidth [Doh36; BMB11].

Similar considerations can be also made for the frequency dependence of the efficiency as shown in Fig. 5.2b. The load impedance is optimized for each frequency point individually, maintaining high PAE over a wide frequency range (A). The contrary scenario is a simple, non-tunable, lossless L-C matching network, designed for 2 GHz and 0 dBm power back-off, resulting in performance degradation, when changing the operation frequency (B). Broadband matching networks are feasible but intrinsically reduce the matching quality and cannot reject the harmonics produced by the PA, if the required bandwidth is to large. Another advantage of tunable impedance matching, is the possibility to flexibly chose between efficient operation or high output power, which will be shown later. Consequently, the average frequency related efficiency $\eta_{\text{eff,f}}$ can be calculated as:

$$\eta_{\text{eff,f}} = \frac{1}{B} \int\limits_{f_1}^{f_2} \text{PAE}(f) \, df \ , \tag{5.3}$$

with the total bandwidth $B = f_2 - f_1$. In this simple case, the overall improvement would result in $\Delta\eta_{\text{eff,f}} = 28\,\%$. Both methods can be combined to a possible figure of merit as a total efficiency:

$$\eta = \frac{1}{B} \int\limits_{f_1}^{f_2} \int\limits_{P_s=0}^{P_s=\infty} \text{PDF}(P_s) \cdot \text{PAE}(P_s,f) \, dP_s \, df \ . \tag{5.4}$$

In addition, tunable impedance matching networks, placed at the input of a transistor, allow the extension of operation frequency, compared to non-tunable input

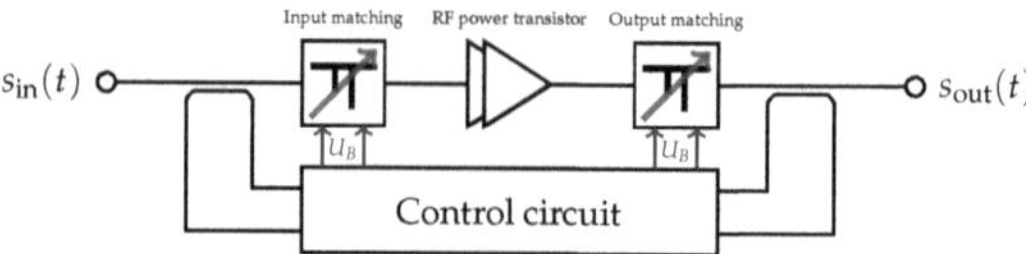

Figure 5.3: Schematic of a transistor with tunable impedance matching at the input and at the output.

matching networks. This further allows to intentionally detune the input impedance, hence, reduce the input power and protect the transistor in case of undesired operation scenarios, e.g. strong reflexions at the load side [Fer+16].

Fig. 5.3 shows a possible implementation of tunable matching networks at the input and at the output of the transistor, aiming for efficiency and output power improvements or VSWR protection. The couplers at the input and output are used to provide feedback for the bias voltage control unit of matching networks.

The implementation of tunable impedance matching at the output of the transistor is a challenging task, as the components are facing large RF power with large RF voltage swings. Therefore, within the scope of this work, the focus is put on the investigation and implementation of tunable output impedance matching networks. A number of technologies are available to implement tunable elements of such a circuit. However, varactor diodes suffer from intrinsically low linearity and limited power handling capability for GHz RF applications and, hence, have to be implemented in large arrays [Raa11]. For the implementation of tunable impedance matching networks, where significantly lower RF power levels can be expected, tunable varactor diodes can be considered as a good alternative. MEMS offer good linearity and moderate power handling capability but suffer from mechanical fatigue [DSB08]. The fabrication of continuously tunable MEMS varactors is also challenging.

Continuously tunable ferroelectric varactors, investigated within this work have shown good tunability, high linearity and high power handling capability, see Chapter 4, and are the key elements for tunable impedance matching networks investigated in the following sections.

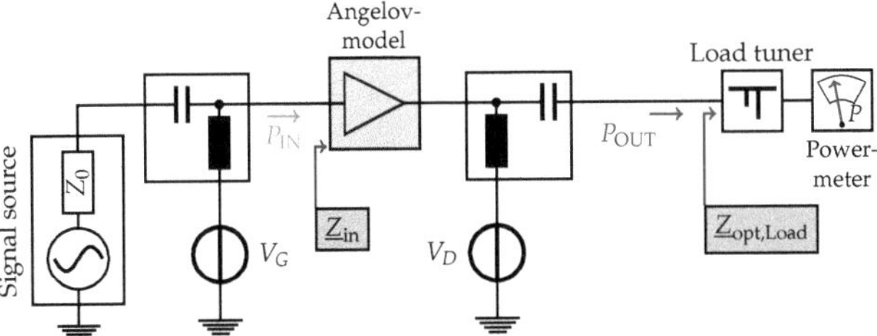

Figure 5.4: Schematic of the harmonic balance simulation in ADS for impedance extraction of the GaN HEMT.

5.1 Transistor Characterization for Impedance Matching

Within the scope of this work, gallium nitride (GaN) high electron mobility transistors (HEMT) from Ferdinand-Braun-Institut (FBH) have been used for hybrid implementation of tunable impedance matching networks. The source and load impedance of such a device is strongly depending on numerous parameters, e.g. bias point, frequency, input power, temperature and many other. Measurement-based extraction of the impedance with these many degrees of freedom is pointless. Therefore, an Angelov model[Ang+05], provided by the FBH was used for the extraction of the optimal load impedance for different input power levels and different frequencies. Fig. 5.4 shows the schematic of a harmonic balance simulation set up in ADS. A single transistor cell with $8 \cdot 250\,\mu\text{m}$ gate width was characterized in class AB ($I_{qs} \approx 80\,\text{mA}$), which is a good compromise between linearity and efficiency. For each frequency and input power level, the variable load was swept and impedance values yielding maximum output power and maximum efficiency were extracted. The input impedance was calculated from the voltage and current at the input. Consequently, at least one iteration was used to extract the required load and source impedance.

Fig. 5.5 shows the input and output impedance dependence over frequency f and input power P_{in} of a single GaN HEMT cell, extracted with the Angelov model in a harmonic balance simulation.

The optimal load impedance $\underline{Z}_{\text{opt,Load}}$ varies over frequency due to the channel capacitance C_{ds} as well as C_{gd}. With increasing input power level the impedance also varies due to thermal effects in combination with rising RF voltage swing

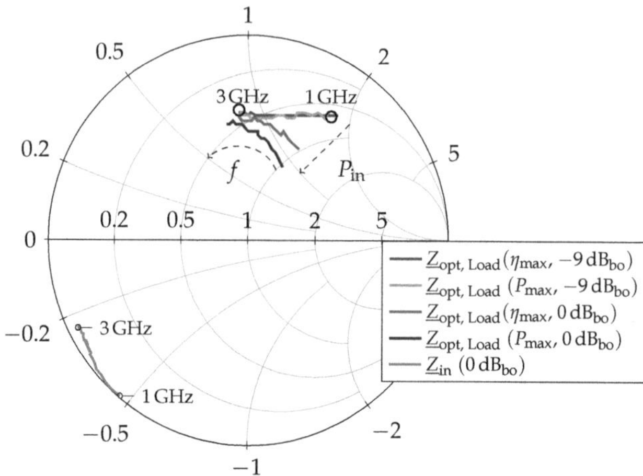

Figure 5.5: Simulated optimal load impedance $\underline{Z}_{\text{opt,Load}}$ and transistor input impedance $\underline{Z}_{\text{in}}$ for different RF power levels and frequencies.

(capacitance variations). For small signal CW analysis it is instructive to model the transistor with an equivalent circuit. The transistor, the generator and the impedance matching network between the generator and the transistor can be modeled as an equivalent source with a source resistor R_s and a shunt capacitor C_s. Here, the capacitor models the frequency dependence, while the resistor models the input power dependence. Fig. 5.6 depicts the transformation, which will be later required for the synthesis of the matching network.

The input impedance $\underline{Z}_{\text{in}}$ of the transistor cell shows mainly frequency related variations due to the C_{gs} capacitance. For further investigations, the transistor is assumed to be fully matched at the input.

In a similar manner, the impedance values for three- and five parallel GaN HEMT cells were obtained.

Within the scope of this work, two different approaches to synthesize a tunable matching network have been investigated and implemented: a CAD approach with a definition of a cost function in an impedance space and a classical filter theory approach, where non-tunable capacitors were replaced with varactors. The latter has the advantage of an inherent frequency matching to a certain extent within

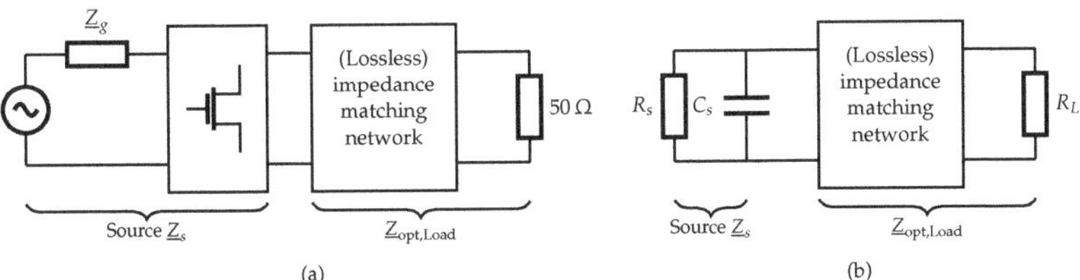

Figure 5.6: (a) Schematic of a complex generator impedance $\underline{Z}_g$ (already ensuring matched condition with the transistor), transistor, matching network and a load. (b) Equivalent representation of the input-matched transistor. The source symbol is omitted for convenience. Note that for power matching $\underline{Z}_s \overset{!}{=} \underline{Z}^*_{\text{opt,Load}}$.

the limits of Bode-Fano[Bod55; Fan48], depending on the required bandwidth. However, the networks derived with the filter theory make use of ideal, lumped elements and, hence, often lead to deviations with realized structures due to real elements' loss and their parasitics. This requires iterative corrections to the circuit, especially when utilizing tunable capacitors. The CAD-aided approach already considers the parasitics and, in addition, allows optimization in more than one dimension (frequency, input power, output power, PAE, ..).

For both synthesis methods, a lowpass impedance matching topology will be considered, having a number of advantages and disadvantages:

- The drain voltage can be provided through the matching network. This has the advantage that no additional bias circuitry is required, which is beneficial for both, measurement and a packaged tunable transistor (seen as a component which has to be deployed in a circuit and with regard to the available space in a transistor's package).

- The harmonics are intrinsically rejected by the lowpass topology. Highpass networks are self-evidentially not expedient.

- Lowpass topology *wastes* some of the matching quality, as it is intrinsically provides good matching condition from DC to the designed cutoff frequency. A bandpass characteristic would be more beneficial with this regard, but would require an additional drain voltage biasing circuitry, which should be avoided. This issue will be addressed by numerical optimization.

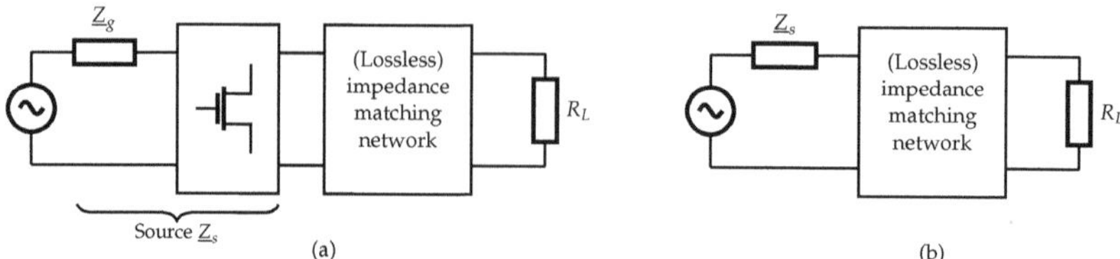

Figure 5.7: Idealized representation of the matching problem (a). Summarized representation by absorbing the transistor into the generator (b).

5.2 Filter-based Matching Network Synthesis

Broadband matching of frequency-dependent impedances has been mathematically treated for non-tunable components and solutions to this problem are known [Bod55; Fan48; CJ79]. However, this mathematical treatment does not cover tunable capacitors and/or inductors. Further, the input power dependent variations of the generator impedance cannot be covered by means of classical filter theory methods. This section covers the investigation on the design approach of *tunable* impedance matching networks, utilizing the methods of filter theory.

Figure 5.7(a) shows the idealized case of a generator with complex source impedance $\underline{Z}_g$ to ensure matched condition with a transistor, followed by a transistor, a lossless matching network and a constant load. The combination of the generator and the transistor can be summarized as an independent generator with a source impedance $\underline{Z}_s$, as indicated in Fig. 5.7(b).

Darlington [Dar39] has shown that any physically realizable arbitrary load (or source) impedance can be approximated as a two-port network terminated by a pure resistance. This resistance can be transformed to any required value incorporating ideal transformers. Consequently, the network shown in Fig. 5.7(b) transforms into a cascade of networks as shown in Fig. 5.8. In this form it becomes apparent, that a matching network can be seen as a part of a network N, which itself can be considered as a filter between two real ports. In a sense, the reactive network N′ partially predetermines the elements of the filter and, given a required framework for the filter, the task is to determine the remaining elements of the network N″. However, first an approximation of the complex source impedance $\underline{Z}_s$ is required.

In the previous section, the optimal load impedances of the transistor cell have been extracted for various operation conditions, which corresponds to the reflexion coefficient Γ_N in Fig. 5.8. Consequently, its complex conjugated can be considered

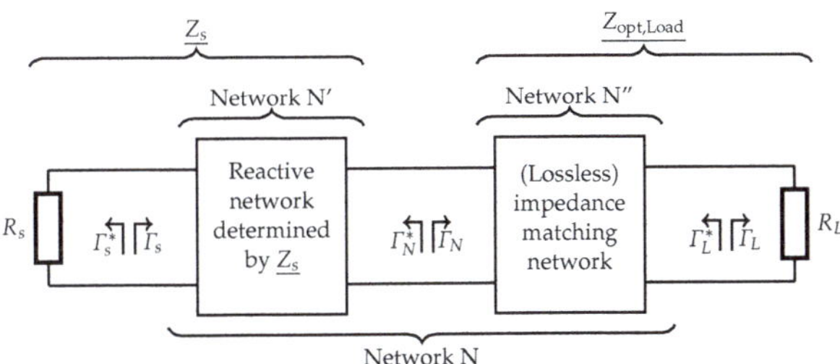

Figure 5.8: Formulation of complex source impedance $\underline{Z}_s$ as a reactive network N′ and a resistance R_s in accordance to Darlington.

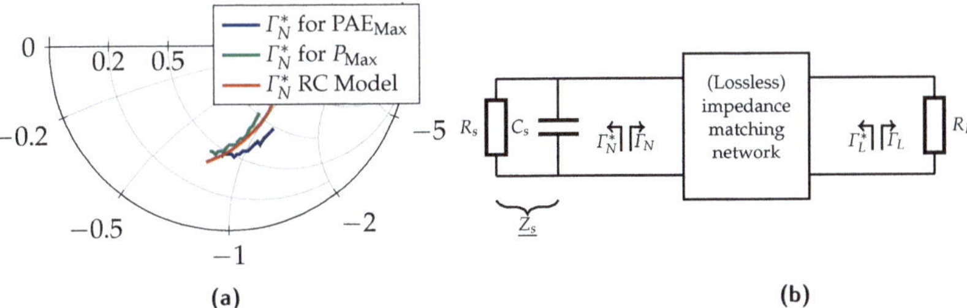

(a)   **(b)**

Figure 5.9: (a) Comparison of Γ_N^* of the HEMT cell extracted from the harmonic balance simulation and the simplified RC equivalent circuit. (b) The simplified equivalent circuit of the HEMT impedance for matching. For the plot, $R_s = 95\,\Omega$ and $C_s = 1.15\,\text{pF}$ were used.

as the complex source impedance $\underline{Z}_s$. In the simplest case, this complex source impedance can be modeled as a resistance R_s with a shunt capacitance C_s. Figure 5.9a exemplary compares the modeled source with the values extracted from the harmonic balance simulation of the transistor from Section 5.1. A compromise between the impedance values yielding maximum efficiency and maximum output power has been chosen. This results in $R_s = 95\,\Omega$ and $C_s = 1.15\,\text{pF}$.

Considering the situation shown in Fig. 5.9, Bode [Bod55][1] has shown a theoretical limitation exists on broadband impedance matching with fixed, lossless, lumped elements given by:

---

[1] Later Fano [Fan48] has generalized this statement for arbitrary loads/sources that can be represented in a ladder structure consisting of series and shunt capacitors and inductors.

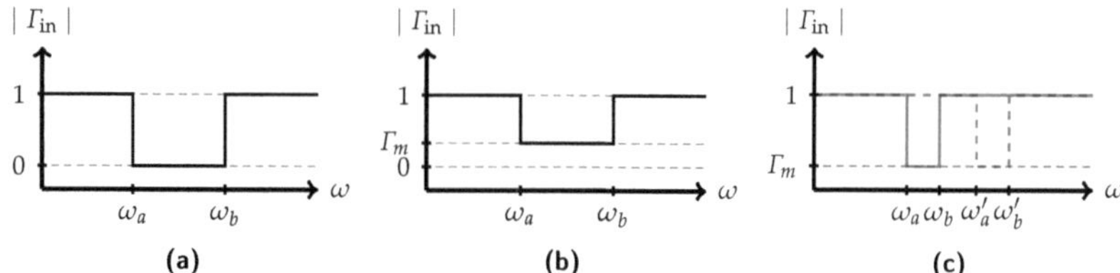

Figure 5.10: (a) Ideal bandpass behavior of the matching network within a specified band $\Delta\omega$. (b) The bandpass behavior according to Bode. (c) Tunable transmission characteristics with reduced bandwidth $\Delta\omega$.

$$\int_0^\infty \ln\left(\frac{1}{|\Gamma_{\text{in}}(\omega)|}\right) d\omega \leq \frac{\pi}{R_L C_L} \ , \tag{5.5}$$

where $\Gamma_{\text{in}}(\omega)$ is the reflexion coefficient seen looking into the arbitrary lossless impedance matching network. This theoretical limit is usually referred to as the Bode-Fano limit. For the case here, it is more convenient to formulate the problem from the load side Γ_L. Aiming ideal transmission within a certain specified band $\Delta\omega$ as shown in Fig. 5.10(a), equation (5.5) yields:

$$\int_{\Delta\omega} \ln\left(\frac{1}{|\Gamma_L^*(\omega)|}\right) d\omega = \Delta\omega \ln \frac{1}{|\Gamma_m|} \leq \frac{\pi}{R_s C_s} \ , \tag{5.6}$$

where Γ_m can be written as:

$$|\Gamma_m| = \exp \frac{-\pi}{R_s C_s \Delta\omega} \ . \tag{5.7}$$

This leads to the following conclusions for the matching problem:

- For a given load (here $R_s C_s$), a matching over a broader bandwidth $\Delta\omega$ can only be achieved at the expense of a higher reflexion coefficient Γ_m, see Fig. 5.10(b). This implies, that the choice of a lowpass topology is *wasting* some of the matching quality. However, the benefits of the lowpass filter prototype have been mentioned in the previous section.

- Γ_m cannot be zero unless $\Delta\omega$ becomes zero. Hence, perfect match is only possible for a finite number of frequencies.

- As R_s and/or C_s increases, the quality of the match ($\Delta\omega$ and/or Γ_m must decrease. Hence, high-Q load/source is harder to match (over a broad frequency range or lower Γ_m) than loads/sources with lower Q.

The use of tunable elements, in a sense, allows to overcome these limitations. A reduction of the transmission bandwidth allows to reduce Γ_m while the tunability of the circuit allows to shift the desired transmission frequency as depicted in Fig. 5.10(c). Further, this approach allows to reject harmonics, which is of great importance for efficient transistor operation [PZA14]. First, a mathematical method to synthesize the remaining, yet not tunable part (N") of the network N is required, which will be the focus of the following investigations.

As has been previously mentioned, the problem can be formulated as a filter between two ports, both having real impedance. The perfect filter would have zero insertion loss in the passband ($|\Gamma| = 0$), infinite attenuation in the stopband ($|\Gamma| = 1$) and ideally a linear phase response to maintain the signal quality. Still, as has been shown for the reflexion, compromises have to be made in the design. The *insertion loss method* gives a high degree of control over the pass- and stopband filter characteristics as well as the phase characteristics, with a systematic way to synthesize the desired response. It is defined by the power loss ratio:

$$P_{\text{LR}} = \frac{\text{Input Power}}{\text{Power delivered to load}} = \frac{P_{\text{avs}}}{P_{\text{load}}} = \frac{1}{1 - |\Gamma(\omega)|^2} \ . \tag{5.8}$$

To obtain a sharper cutoff in the stopband, the insertion loss characteristic is modeled with the Chebyshev polynomials [Poz05]:

$$P_{\text{LR}} = 1 + k^2 T_N^2 \left(\frac{\omega}{\omega_c} \right) \ , \tag{5.9}$$

where T_N are the Chebyshev polynomials defined as:

$$T_N(x) = 2x T_{N-1}(x) - T_{N-2}(x), \text{with } T_1(x) = x \ . \tag{5.10}$$

Or more generally

$$T_N(x) = \begin{cases} \cos(N \cos^{-1} x), & |x| \leq 1 \\ \cosh(N \cosh^{-1} x), & |x| > 1 \end{cases} . \tag{5.11}$$

The polynomials $T_N(x)$ oscillate between -1 and +1 for $|x| \leq 1$, as shown in Fig. 5.11(a), while the power loss ratio cannot be less than 1. Therefore, it has to be mapped and scaled, which leads to the expression in (5.9). Further, $x = \omega/\omega_c$

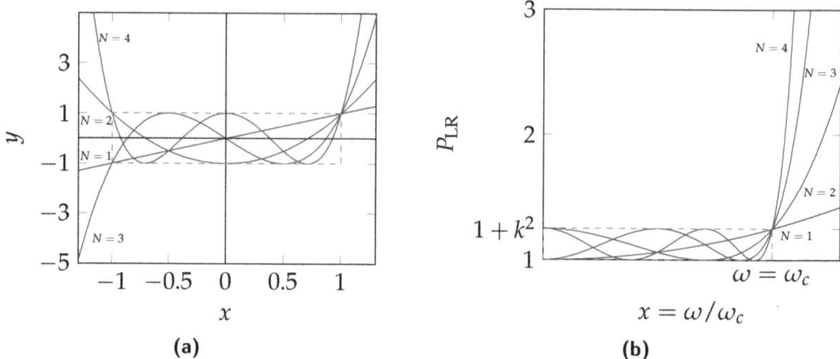

Figure 5.11: (a) Chebyshev polynoms up to the 4th order. (b) Mapping to power loss ratio.

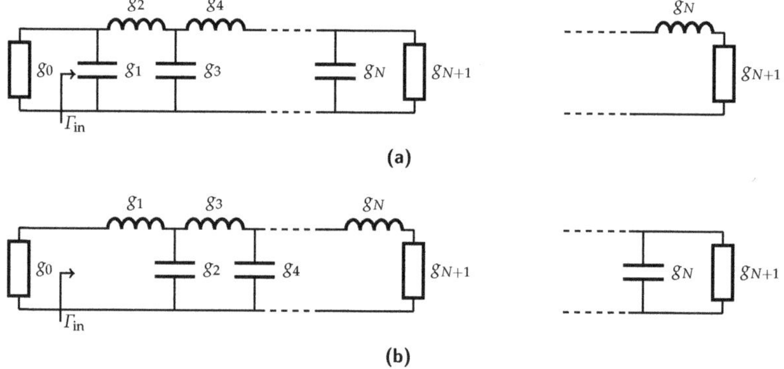

Figure 5.12: Prototype ladder filter functions with normalized element values g.

is mapped to ensure $|x| \leq 1$ within the passband. In this context, the expression k^2 determines the ripple level within the passband, see Fig. 5.11(b).

The prototype filter topologies of interest for the presented problem are depicted in Fig. 5.12. The lowpass topology allows to provide the transistor drain-voltage through the network. Further, the lowpass structure intrinsically rejects high frequencies, which is beneficial for the transistor in terms of harmonics. The normalized element values g are calculated depending on the filter order N, the ripple level k^2 as well as the cutoff frequency ω_c by calculating the reflexion coefficient of the prototype network Γ_{in} and solving it in combination with (5.9). The normalized values can be calculated as [MYJ63]:

$$
\begin{aligned}
g_0 &= 1 \\
g_1 &= \frac{2a_1}{\gamma} \\
&\;\;\vdots \\
g_k &= \frac{4a_{i-1}a_i}{b_{i-1}g_{i-1}} \qquad , i = \{2,..,N\} \\
g_{N+1} &= \begin{cases} 1 & , N \text{ is odd} \\ \coth^2\left(\frac{\beta}{4}\right) & , N \text{ is even} \end{cases}
\end{aligned}
\tag{5.12}
$$

with

$$
\begin{aligned}
\beta &= \ln\left[\coth\left(\frac{L_r}{17.37}\right)\right] \\
\gamma &= \sinh\left(\frac{\beta}{2N}\right) \\
a_i &= \sin\left(\frac{\pi(2i-1)}{2N}\right) \qquad , i = \{1,..,N\} \\
b_i &= \gamma^2 + \sin^2\left(\frac{i\pi}{N}\right) \qquad , i = \{1,..,N\} \quad .
\end{aligned}
\tag{5.13}
$$

Here L_r it the ripple level k^2 in decibel ($L_r = 10\log[1 + k^2]$). For various combinations, the results are summarized in [MYJ63]. An excerpt is given in Table D.1. The de-normalized values can be calculated as [Poz05]:

$$
\begin{aligned}
R_s &= g_0 R_0 \\
R_L &= g_{N+1} R_0 \\
L &= \frac{R_0 g_L}{\omega_c} \\
C &= \frac{g_C}{R_0 \omega_c} \quad .
\end{aligned}
\tag{5.14}
$$

Here, g_C and g_L refer to the normalized element values of inductance L and capacitance C and R_0 is the required port impedance. An important consequence of the approach is that any **even** filter order N leads to a transformation between g_0 and

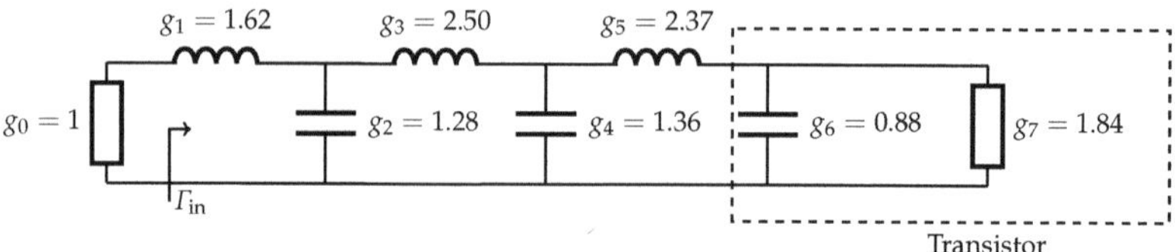

Figure 5.13: Prototype lowpass filter structure with $N = 6$ and $L_r = 0.4\,\text{dB}$. Note that for convenience reasons Γ_{in} corresponds to Γ_L^* in Fig. 5.9(b). So the circuit can be considered as mirrored.

g_{N+1} depending on the ripple level, while the odd filter order always results in equal term impedances g_0 and g_{N+1}, see (5.12). This property will be later used for the network design.

A filter prototype function with $N = 5$ elements, shown in Fig. 5.12(b), is of particular interest. It comprises:

- Two shunt capacitors, which can be later replaced with varactors. This is the minimum number of varactors required to vary Γ along two dimensions (real and imaginary axis).

- The network starts and ends with inductances, which can be (partially) realized as bond wires for later implementation of the network with a transistor cell. So parasitics of the wire-bonds can be absorbed into the actual network.

However, a Chebyshev filter with the order $N = 5$ intrinsically has the same port impedance $g_0 = g_{N+1} = 1$, which are both purely real according to (5.12), while the transistor requires a transformation from a complex source ($R_s = 95\,\Omega$, $C_s = 1.15\,\text{pF}$). One possible approach is to synthesize a filter with an even order $N = 6$ with a predefined ripple level to receive the required transformation. In this case, $L_r \approx 0.4$ yields to the filter structure shown in Fig. 5.13 with the required transformation between g_0 and g_{N+1}. Note that the prototype shown in Fig. 5.13 is mirrored, compared to Fig. 5.9(b) for convenience reasons. Consequently, the matching problem is now formulated from $\Gamma_{\text{in}} \equiv \Gamma_L^*$ instead of Γ_N.

Ideally, the last shunt capacitance should be designed in a way that it can be completely *absorbed* into the load ($C_6 \overset{!}{=} 1.15\,\text{pF}$), which in accordance to (5.14) would

Table 5.1: Obtained non-tunable matching network element values.

R_s	L_1	C_2	L_3	C_4	L_5	C_6	R_L
50 Ω	4.85 nH	1.54 pF	7.48 nH	1.62 pF	7.08 nH	1.15 pF	95 Ω

lead to $\omega_c = 2\pi \cdot 2.66\,\text{GHz}$. This value is used to determine the de-normalized values as a starting point for further optimization, which are summarized in Table 5.1[2].

To introduce tunability, the capacitors C_2 and C_4 are replaced with varactors. Fig. 5.14(a) shows the variation of the insertion loss with the capacitance values swept between 0.5 pF and 10 pF, while Fig. 5.14(b) shows the impact on the cutoff frequency of the lowpass filter. Depending on the tolerable insertion loss, a range for the cutoff frequency can be defined. The dashed area in Fig. 5.14(b) shows the variation of the cutoff frequency under the condition that the insertion loss is less than IL $\leq 0.01\,\text{dB}$. With the approximate tunability $\tau \approx 50\,\%$ of thick film MIM varactors, the capacitance values C_2 and C_4 can be defined for a required tuning range of the cutoff frequency. For a specified range of 1 GHz..3 GHz, the required values are $C_2 = 1.5\,\text{pF}..3\,\text{pF}$ and $C_4 = 1.5\,\text{pF}..3\,\text{pF}$, see Fig. 5.14(b). The available tunability of $\tau \approx 50\,\%$ does not allow to fully cover the targeted frequency range. The response of the insertion loss for various tuning states is summarized in Fig. 5.15. With a harmonic rejection better than 10 dB the lower operation frequency of the network can be defined as $f_l = 1.15\,\text{GHz}$, while the transmission at highest frequency gives the upper operation frequency with $f_u = 2.7\,\text{GHz}$.

This investigation, however, does not include the parasitics and losses of the elements, which cannot be avoided in physical implementation of the matching network. Fig. 5.15(b) shows the impact on the insertion loss with an estimated quality factor $Q_C = 50$ and $Q_L = 100$ for capacitors and inductors respectively (except C_6). The values C_2 and C_4 were chosen in accordance to the lossless case. Three major aspects can be observed:

- The ripple level has an offset, which can be attributed to the energy dissipated in the elements.

- The insertion loss does not longer follow the Chebyshev characteristic even with C_2 and C_4 according to the element values derived in Table 5.1. This can

---

[2]It has to be noted, that this approach is only applicable for positive transformation ratios (here 50 Ω to 95 Ω), which is also defining the ripple level L_r. Further, the capacitance C_6 not necessarily leads to an appropriate cutoff frequency. These cases are treated in the Master thesis of Felix Lenze [Len16].

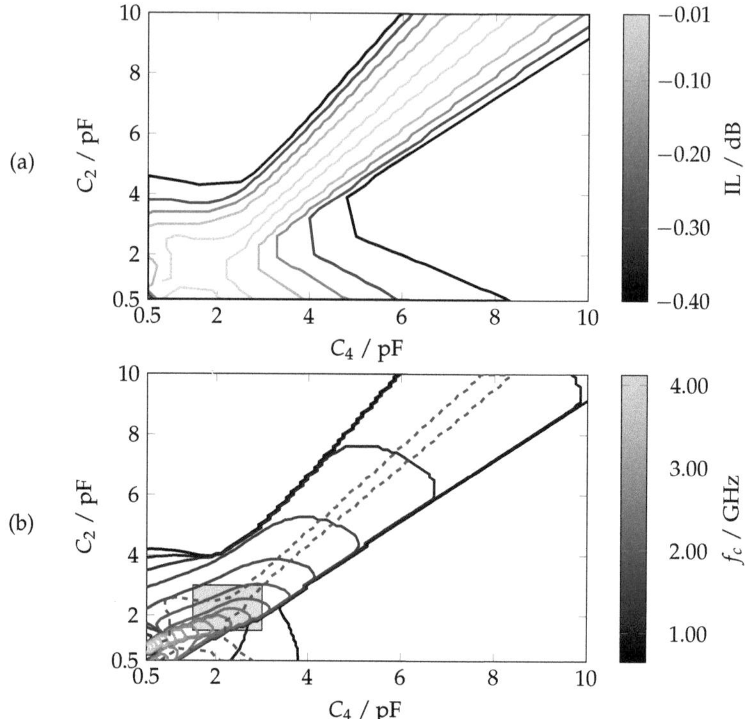

Figure 5.14: Contour plot of the insertion loss (a) and cutoff frequency (b) versus the change in capacitance values C_2 and C_4. The dashed lines in (b) give the possible tuning under the condition that the insertion loss is below 0.01 dB. The shaded area gives the optimum values for C_2 and C_4 to achieve maximum tuning of the cutoff frequency.

be attributed to shift in resonance frequencies as a consequence of resistive loading.

- Assuming a maximum tollerable inertion loss of IL = 0.8 dB the upper cutoff frequency reduces down to $f_u = 2.1$ GHz. With a harmonic suppression of at least 10 dB the lower frequency f_l stays at around 1.1 GHz.

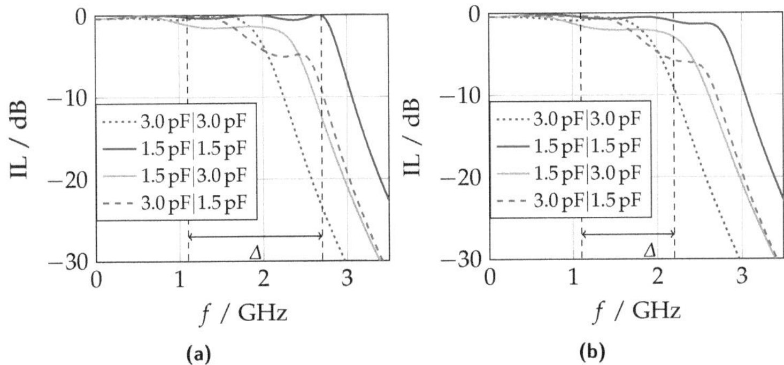

Figure 5.15: The insertion loss of the lossless (a) and lossy (b) matching network for different configurations of capacitors C_2 and C_4. The $Q_C = 50$ and $Q_L = 100$ were assumed in (b) for the capacitors and the inductors, respectively.

These impacts can be considered within the choice of the capacitance values C_2 and C_4 in a similar manner as shown in Fig. 5.14 by plotting the insertion loss contours in combination with the cutoff frequency under the assumption of lossy components, see Fig. C.1 in Appendix C. Further, the element values of L_1, L_2 and L_3 are not necessarily the best choice when coping the losses and have to be numerically corrected.

However, the specified frequency range is 1 GHz to 3 GHz. Equation (5.6) shows, that the quality of matching Γ_m and the bandwidth are inter-dependent. Consequently, the matched frequency range below 1 GHz unnecessarily reduces the quality of the matching. The element values are therefore adjusted to reduce the matching quality in the lower frequency range. The result of the optimization is shown in Fig. 5.16 with the corresponding element values summarized in Table 5.2. Note that the matching around 1 GHz is reduced.

It becomes apparent that the filter theory approach to design a tunable impedance matching network can be considered as a mere starting point for further optimiza-

Table 5.2: Numerically optimized element values to take into account the quality factor of elements and to reduce the matching quality below 1 GHz.

R_s	L_1	C_2	L_3	C_4	L_5	C_6	R_L
50 Ω	1.2 nH	3.3 pF	3.7 nH	4.0 pF	5.0 nH	1.15 pF	95 Ω

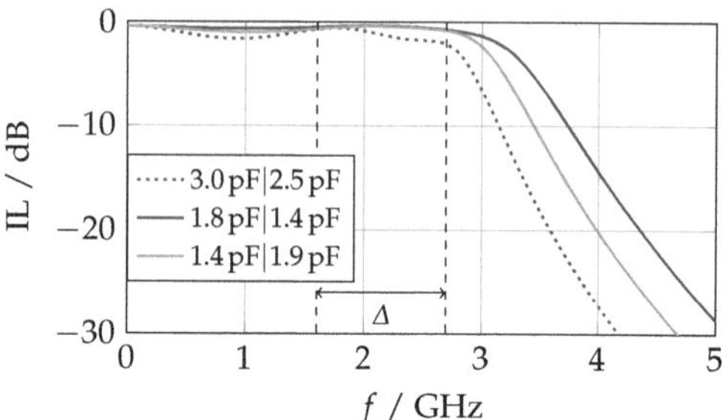

Figure 5.16: The insertion loss of the tunable lossy matching network after numerical optimization for different configurations of capacitors C_2 and C_4. The $Q_C = 50$ and $Q_L = 100$ were assumed for the capacitors and the inductors respectively. Note the reduced matching quality for lower frequencies compared to Fig. 5.15(b).

tion. The parasitic effects of real components deteriorate the synthesized network in a way that numerical corrections to the obtained circuit elements values are inevitable. Still, this method reduces the required optimization space to a manageable extent.

5.2.1 Thick Film Implementation

The filter-based network synthesis strongly relies on reactances without parasitics. Therefore, to minimize the parasitics of the elements, the network was implemented in MIM thick film technology in combination with commercial SMD inductors. A compact circuit board with a total size of 2.5 mm × 0.9 mm was processed on AlN carrier substrate in CPW configuration to allow simple on-wafer characterization. The fabrication was subdivided into three steps.

1. Lithographic fabrication of the bottom gold electrode in accordance to the fabrication described in Sec. 2.3.2.

2. Screen printing of the BST layer followed by sintering at 850 °C for 1 hour.

3. Lithographic fabrication of the top gold electrode in accordance to the fabrication described in Sec. 2.3.2.

The fabricated circuit board exhibits a bottom electrode height of approximately 3.5 μm. A low temperature sintered Cu-F co-doped BST composite (with 5 % of ZnO/H_3BO_3 sinter additive) with an approximate relative permittivity of $\varepsilon_r \approx 200$ was used for the work, see [Koh16]. A BST layer height of $h_{BST} \approx 4$ μm was assumed in the design of the MIM varactors to achieve the required capacitance values for C_4 and C_2. To obtain higher power handling capability and to have an isolated bias electrode, the varactors were implemented as double capacitors and pairwise in shunt configuration. The area of individual single capacitors is:

$$A(C_4) = 50 \cdot 200\text{μm}^2$$
$$A(C_2) = 35 \cdot 200\text{μm}^2$$

(5.15)

In accordance to the results found in Section 4.2 (4.8), the varactor C_2 and C_4 would sustain at least 3.2 W and 4 W, respectively. However, the rated power was derived for series configuration. Here, the varactors are placed in parallel, which reduces the RF power seen by the varactors and, hence, increases the RF power that the matching network can sustain.

The top electrode thickness is approximately 2.5 μm. The fully processed circuit board is shown in Fig. 5.17(a).

After the fabrication, the circuit board with implemented thick film varactors was equipped with 0201 SMD inductors (TDK MHQ0603P Series). The fully assembled circuit along with its equivalent circuit is depicted in Fig. 5.17(b).

For small signal characterization, the network is designed to be contacted with 150 μm-pitch GSG probes. Large contact pads at the input and at the output (red-shaded areas in Fig. 5.17(b)) allow hybrid implementation with a transistor cell by means of wire bonding. Four pin-probes are used to apply bias voltage to the varactors, which are marked with green-shaded areas in Fig. 5.17(b). Bias-tees at the input and at the output were used to apply reference potential for the varactors. The varactor bias voltage was swept between 0 V and 200 V and the response of the circuit was measured from 100 MHz to 6.1 GHz in a 50 Ω-system.

The measured S-parameter data was re-normalized in ADS to estimate the performance of the matching network in combination with the transistor cell, since it is intrinsically designed for non-50 Ω environment. Further, a harmonic balance simulation of the measured S-parameters in combination with the Angelov model of the transistor cell was carried out to estimate the PAE and P_{out} of the module. The results are summarized in Fig. 5.18. For each frequency point, the varactor bias voltage yielding minimum return loss was extracted with the corresponding

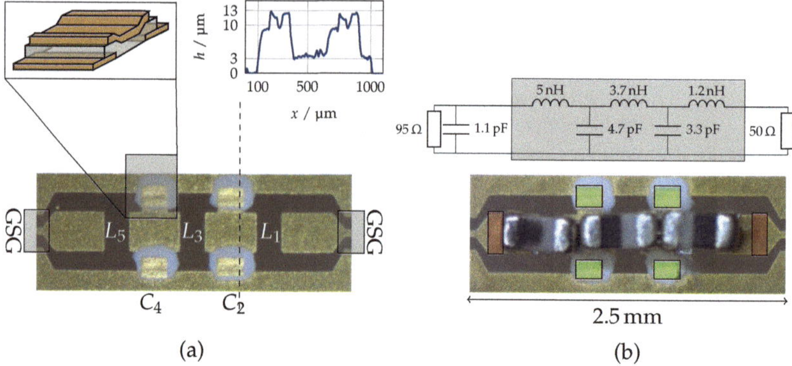

Figure 5.17: (a) Topview on the fabricated tunable matching network implementing double MIM varactors, see inset, as well as the cross-section view through two shunt double MIM varactors. (b) SMD-equipped tunable matching network and the equivalent circuit model. The red-shaded areas denote place for bonding, while the green-shaded areas mark the points for varactor biasing. Note that the element values are now mirrored compared to Fig. 5.13.

insertion loss. Fig. 5.18(a) shows the envelope of this analysis. The dashed line in Fig. 5.18(a) shows a fix bias voltage configuration over frequency to demonstrate the tuning. Especially around 2 GHz a significant reduction of the return loss can be achieved. Though the matching is better than 10 dB throughout the entire frequency range, the average insertion loss higher than 0.8 dB (typically $\geq$ 2 dB). The analysis of the fabricated BST thick film varactors revealed a significant deviation between the assumed BST layer thickness (4 µm) and the fabricated thickness of $h_{BST} \approx 6$ µm. Therefore, the fabricated capacitance values are at least 33 % below the specified values, which is one of the reasons for the poor performance.

However, the co-simulation of the measured S-parameters and the Angelov model reveals a drastic drop in PAE and P_{out} as a consequence of high insertion loss. For the evaluation in Fig. 5.18(b), for each frequency point the varactor bias configuration yielding maximum PAE with the corresponding P_{out} was chosen and vice versa. In a sense it shows the envelope for all measured varactor bias states. An influence of the varactors between the PAE and P_{out} can only be observed for frequencies below

114

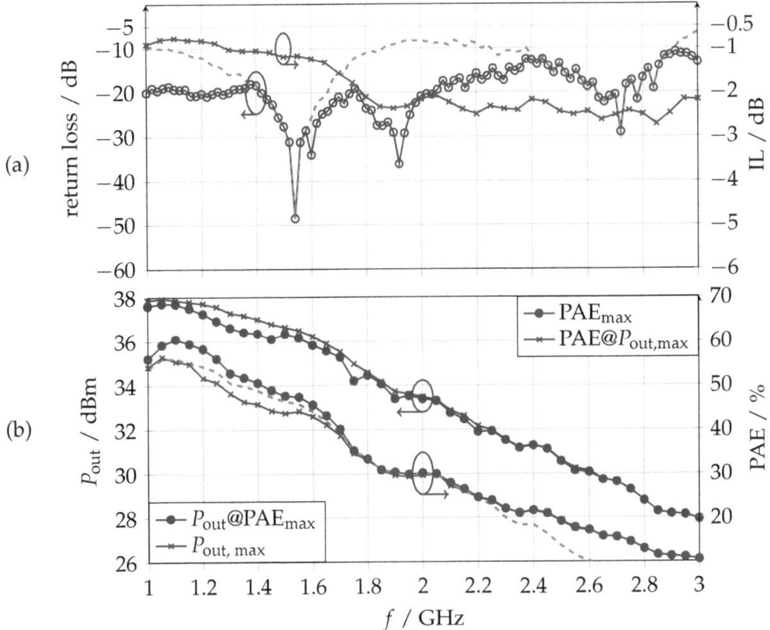

Figure 5.18: (a) The envelope of minimum return loss of the tunable matching network in combination with the transistor cell with the corresponding insertion loss. (b) Result of the harmonic balance simulation of the HEMT model and the measured S-parameter to estimate PAE and P_{out} of the module. At each frequency point, the curves show configuration yielding maximum PAE with the corresponding P_{out} and vice versa. The dashed lines show a fix varactor tuning state versus frequency.

2 GHz with a maximum difference around 1.4 GHz, yielding a ΔPAE $\approx 5\%$ and $\Delta P_{out} \approx 0.8$ dB. The dashed line shows the PAE behavior for a fix varactor tuning state, optimized for maximum PAE at 2 GHz. A minor increase throughout the entire band can be observed. Surprisingly, almost no improvement can be observed in PAE in the band where the return loss can be significantly varied (1.6 GHz to 2.4 GHz). This effect can also be attributed to the insertion loss. The dielectric losses appear to be higher than expected, which additionally deteriorates the performance even though the matching is improved. A deeper analysis of the fabricated varactors

revealed a quality factor $Q_C \approx 13$ around 2 GHz. A delamination of the BST layers and the top and bottom electrode has been observed, which can explain the strong reduction in Q_C compared to [Koh16]. This would also lead to further reduction of required capacitance values. The delamination of the metal layers and the carrier substrate also rendered the wire bonding to a transistor cell impossible, which did not allow a real module realization. The bottom electrode – substrate delamination is expected to arise from oxidization of the chromium-nickel adhesion layer during the sintering of the BST layer. Process optimization in this direction is required to render the integration possible.

5.2.2 Thin Film Implementation

The synthesis approach was initially intended to be implemented in thick film technology with MIM varactors described in Sec. 2.3.2. However, the delamination of the bottom electrode after sintering does not allow wire bonding of structures. Further, the fabrication process of the screen-printed BST layer and the resulting tolerances is not optimized, giving significant mismatch between required and realized capacitance values. Therefore, commercial thin-film varactors have been used for the implementation of the matching network.

A compact circuit board with a total size of 5 mm × 2.8 mm was processed on alumina carrier substrate, see Fig. 5.19(a). The structuring of the board was performed in accordance to the fabrication steps described in Sec. 2.3.2. To take into account the skin-effect around 2 GHz ($\delta \approx 1.9\,\mu$m) the metallization layer of the mini PCB was galvanically grown to $h_{\mathrm{cond}} \approx 3\,\mu$m. The gold electrodes allow for wire bonding to the transistor cell, while the alumina substrate ensures sufficient heat transfer. The board was equipped with SMD 0402 inductors (TDK Series MHQ1005P) and SMD BST varactors (STMicroelectronics Parascan^{TM}, [STM15]). The fully equipped circuit board is depicted in Fig. 5.19 with the same annotation as for the thick film version. Note that the network dimensions have also increased due to packaging size of the commercial BST varactors. Therefore, also larger SMD inductor packaging was used for convenience reasons. The circuit was measured in the same manner as the thick film version with the measurement results summarized in Fig. 5.20. The same evaluation as for the thick film matching network has been carried out. A significant reduction in insertion loss can be observed compared to the thick film implementation, which can be attributed to the quality factor of the varactors, as well as proper values for the circuit, see Fig. 5.20(a). The insertion loss of the circuit is below 1 dB up to 2.5 GHz. The gray dashed lines show the return loss of two

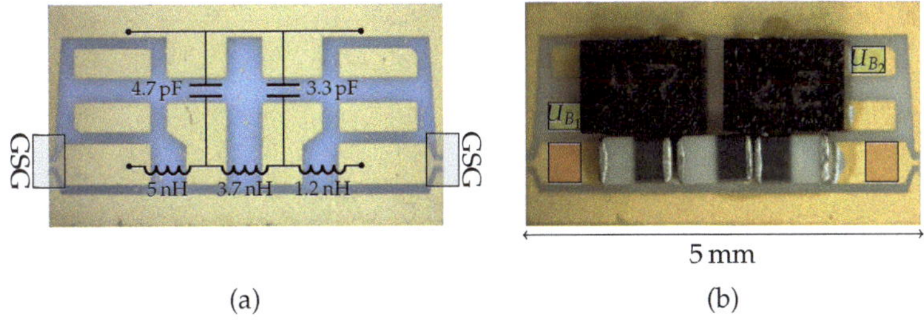

Figure 5.19: (a) Topview on the fabricated miniaturized circuit board with equivalent circuit model. (b) SMD-equipped circuit board. The red-shaded area denote area for bonding while the green-shaded area marks the points for varactor biasing.

random varactor bias states to demonstrate the tuning. The simulation shows negligible impact of varactor tuning between the maximum PAE and the corresponding P_{out} and vice versa, see Fig. 5.20(b). However, it shows a strong influence when tuning over frequency with improvements up to 15 %-points in PAE. This can be observed within the matching bands (1.4 GHz to 2 GHz and 2.4 GHz to 3 GHz). Below 1.4 GHz the performance is strongly reduced, which is caused by insufficient harmonics suppression, compare Fig. 5.16. Nevertheless, even the implementation with commercial elements yields deviation from the designed behavior, when using lossy components with parasitics. Further, the design routine only considers the frequency behavior of the source (or load), neglecting the input power dependence. This can be observed in the fact that almost no improvement between optimal PAE and maximum P_{out} can be achieved. Therefore, a purely numerical circuit design approach is developed and investigated, which includes not only the frequency variations but also the power dependence of the optimal load impedance $\underline{Z}_{opt,Load}$. This synthesis is the focus of the following investigation.

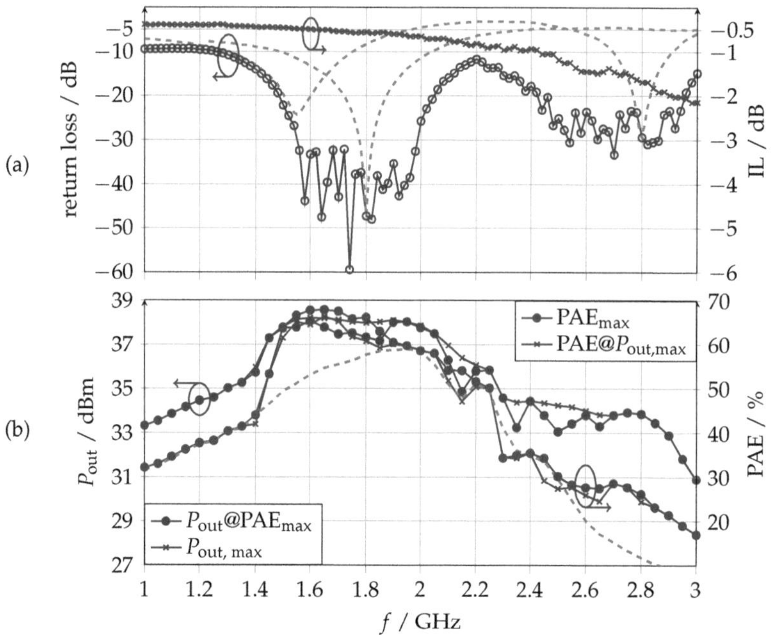

Figure 5.20: (a) The envelope of minimum return loss of the tunable matching network in combination with the transistor cell and the corresponding insertion loss. The dashed lines show the return los of random (but fix) varactor bias states. (b) Result of the harmonic balance simulation of the HEMT model and the measured S-parameter to estimate PAE and P_{out} of the module. Optimized for maximum PAE with the corresponding P_{out} and vice versa. The dashed line indicates a fix tuning state (18 V | 20 V) optimized for 2 GHz.

5.3 CAD-aided Impedance Space Matching Network Synthesis

The prerequisites for the CAD-aided approach are accurate models of components, which should be combined into a circuit. The approach was tested with inter digital capacitors (IDC). The model of IDC varactors presented in Section 3.2.1 has been

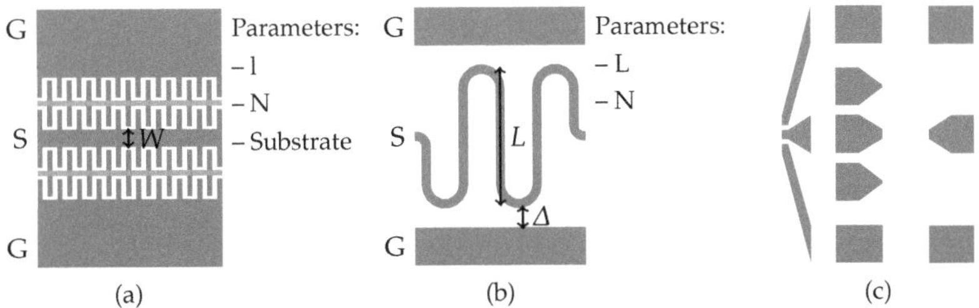

Figure 5.21: Implemented layout components in ADS. (a) IDC varactor with variable finger gap g, width w, length l and number N. Tunability is considered by varying the substrate. (b) Meander line with variable number N and length of the windings L. (c) A set of pads and tapers for contacting purpose.

extended to coplanar versions of varactor pairs, see Fig. 5.21(a). For the sake of simplicity, inductances were implemented as meander lines in CPW configuration, see Fig. 5.21(b). Further, a set of tapers and pads were implemented as individual components, see Fig. 5.21(c). Each of the components was simulated individually. The gap and width of the IDC varactors were kept constant at $g = w = 15\,\mu\text{m}$. The signal line was set to a width $W = 100\,\mu\text{m}$ with regard to required power handling capability. The length l of the fingers was swept from $50\,\mu\text{m}$ to $400\,\mu\text{m}$ in $10\,\mu\text{m}$ steps. The number N of fingers was swept from 2 to 20 in steps of one. This gives a total of $36 \times 19 = 684$ geometry variations of the IDC. To include tunability, each geometrical configuration was simulated on 9 substrates with a BST layer permittivity ranging from $\varepsilon_{\text{r,BST}} = 260..100$ in steps of 20. This gives a total number of $36 \times 19 \times 9 = 6156$ variations for the varactors. The permittivity of the BST layer was chosen in accordance to the Fe-F codoped BST which is expected to have the lowest $\tan\delta$ [Zho12].

For compatibility reasons, the meander line length L (including the arc) was swept in a way that the total width (distance from ground to ground of the CPW) covers similar values as achieved by changing the length l of the IDC components ($L = 4l + 4g + 2w + W - 2\Delta$). To have a fix distance of the arc and the ground lines, the variable $\Delta = 50\,\mu\text{m}$ was used. The number of windings N was swept from 1 to 10 in steps of one. The meander lines are not tunable, therefore, the geometries were simulated on a single, untuned substrate with a BST layer permittivity $\varepsilon_{\text{r,BST}} = 260$. This gives a total number of $36 \times 10 = 360$ different meander line layouts. The pads and tapers shown in Fig. 5.21(c) were simulated with geometry parameters compatible to the IDC/meander line components on an *untuned* substrate.

As has been shown in the previous section, the impedance of the GaN HEMT cell has at least two dependencies, the frequency and the input power level, and is spanning a surface in the impedance space. Therefore, at least two tunable elements are necessary to obtain the required degree of freedom for the tunable matching network. A more convenient representation of this surface can be obtained by a bilinear transformation to a Smithchart.

To keep the insertion loss of the TMN as low as possible, the number of elements was limited to two shunt IDC varactors and a single series meander line, as shown in Fig. 5.22(a). The highlighted elements in Fig. 5.22(a) denote the the components which were simulated on substrates with different permittivity of the BST layer. The input impedance of all individually swept element configurations is shown in a Smithchart above each component. Series varactors were avoided to allow the transistor biasing through the TMN. Fig. 5.22(b) shows the equivalent circuit of the TMN topology with the primary electric properties of each circuit block. The parasitics of the connection pads and tapers are treated as series inductors. One advantage of this approach is that each building block of the circuit contains all parasitics including losses, and hence accurately models the behavior of the resulting circuit[3]. However, one major challenge is the design with a specific characteristic of the network. One could start with the filter synthesis as a starting point for optimization, but the large parasitic inductance of the utilized components would require large corrections to the obtained values. Also, the numerical optimization cannot be avoided, therefore a different approach is followed.

Keysight ADS offers a set of optimization algorithms for a set of pre-defined goals. Most of these algorithms get stuck in a local minimum/maximum when optimizing towards a certain goal. Therefore, the genetic algorithm was employed as the optimizer. It is a heuristic algorithm motivated by nature. It comprises a set of random configurations (*parents*), where the ones with the lowest cost function are combined and considered for next iteration (*children*). In each iteration, the pool is extended with random changes of the children (*mutations*). This way the optimization towards a local minimum/maximum is avoided.

The definition and the formulation of a cost function is essential for this approach. It can be based on PAE or P_{out} optimization, but would require a harmonic balance simulation with a transistor model for each iteration, which would eventually lead to long simulation time. Depending on chosen network elements this approach could even lead to instabilities. In this work, the cost function was defined as the difference in the overlap areas of the impedance space of $Z_{opt,Load}$ for the

---

[3]Filter theory synthesizes a network with idealized lumped elements, therefore realized structures usually differ from simulation.

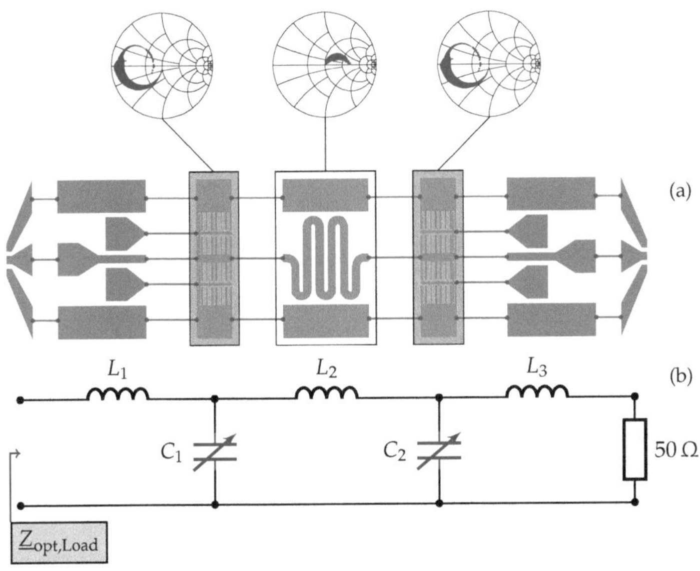

Figure 5.22: (a) Schematic of the π-topology circuit implemented with the building blocks shown in Fig. 5.21. The input impedance of each block for all combinations are depicted in a Smithchart above the elements. (b) Equivalent circuit representation of the layout with the primary electric properties.

transistor's load and the input impedance $\underline{Z}_{in}$ of the tunable impedance matching network. It has to be noted, that the optimizer was equipped with a constraint to ensure compatibility between the building blocks and to avoid unnecessary tapering regions. Further, knowing that the implementation of the TMN with the GaN HEMT cells would require bond wires between the output of the transistor and the TMN, the cost function was defined such, that the inductance of the bond wires finalizes the circuit.

Fig. 5.23(a) shows the required load impedances of three parallel GaN HEMT cells obtained from a harmonic balance simulation at saturated power and at 6 dB power back-off. The optimal load impedance $\underline{Z}_{opt,Load}$ was approximated with the gray-shaded area and was used for the cost function. The circuit, combined with the genetic algorithm, is depicted in Fig. 5.23(b). The design features a total size of 2.8 mm $\times$ 2.7 mm. The input impedance $\underline{Z}_{in}$, obtained with this configuration, is

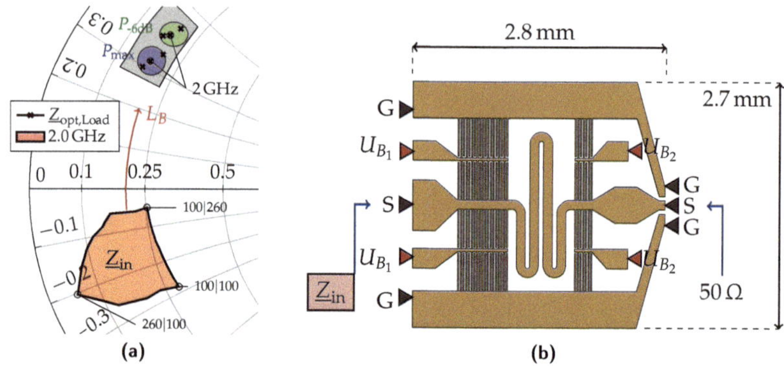

Figure 5.23: (a) The optimal load impedances of three GaN HEMT cells (gray) as well as the input impedance $\underline{Z}_{\mathrm{in}}$ of the circuit combined by the genetic algorithm (red). (b) Layout of the circuit combined by the genetic algorithm. The right sight is tapered for a 150 μm-pitch GSG probe.

shaded red in Fig. 5.23(a). Any arbitrary value within the surface can be obtained by tuning the permittivity of the substrate layer. Here, to estimate the tuning of the input impedance, the values were swept from $\varepsilon_{\mathrm{r,BST}} = 260$ to $\varepsilon_{\mathrm{r,BST}} = 100$. A bond wire inductance $L_B \approx 1.6\,\mathrm{nH}$ shifts $\underline{Z}_{\mathrm{in}}$ towards $\underline{Z}_{\mathrm{opt,Load}}$. The design properties of the presented circuit are summarized in Table 5.3. The primary electrical proper-

Table 5.3: Layout parameters of the circuit in Fig. 5.23(b).

	IDC$_1$	Meander	IDC$_2$
g	15 μm	-	15 μm
w	15 μm	-	15 μm
W	100 μm	100 μm	100 μm
N	10	3	4
l	400 μm	-	400 μm
Δ	-	50 μm	-

ties of individual blocks were extracted and an equivalent circuit of the design is given in Fig. 5.24(a). For the lumped elements representation, the tunability was implemented by a reduction of the values C_1 and C_2 proportional to the permittivity changes of the substrate in the ADS layout simulation.

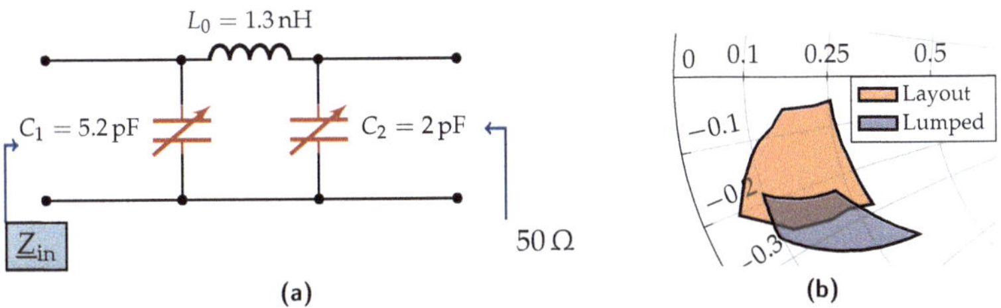

Figure 5.24: (a) Equivalent circuit representation of the layout shown in Fig. 5.23(b). (b) Comparison between the input impedance of the equivalent circuit and the layout.

$$\varepsilon_{\mathrm{BST}} = [260 \cdots 100] \rightarrow \tau \approx 60\,\%$$
$$\rightarrow C_1 = [5.2\,\mathrm{pF} \cdots 2\,\mathrm{pF}]$$
$$\rightarrow C_2 = [2\,\mathrm{pF} \cdots 0.8\,\mathrm{pF}] \tag{5.16}$$

The comparison between the tunable input impedance of the lumped element representation and the layout simulated with ADS is given in Fig. 5.24(b). As can be seen, the parasitics which are considered within the layout model lead to a significantly different result, highlighting the necessity of accurate layout and substrate modeling.

As has been concluded in [Zho12], Fe-F codoped BST exhibits reduced losses compared to undoped and other metal dopings. For high power applications, the reduction of dielectric losses is critical. Therefore, the structure in Fig. 5.23(b) was fabricated on a 635 µm aluminum oxide wafer with screen-printed, Fe-F codoped BST layer of $\approx 6\,\mu\mathrm{m}$ thickness. The thickness varied from the targeted layer thickness of 4 µm due to screen printing tolerances. However, the impact on the IDC capacitance is significantly less than for the case of the MIM varactors.

The structuring of the circuit was performed in accordance to the fabrication of IDC varactors described in Sec. 2.3.2. First, an adhesion layer of 20 nm chromium and 60 nm gold seed layer were evaporated, followed by one-step photo lithography. In a second step, the structure was galvanically heightened to 3 µm, to take into account the skin-effect around 2 GHz ($\delta \approx 1.9\,\mu\mathrm{m}$). Subsequently, the seed layers were etched.

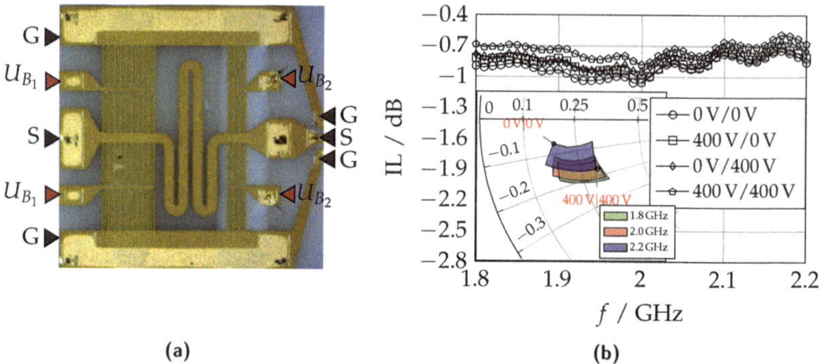

(a)   (b)

Figure 5.25: (a) Microscope image of the fabricated TMN. (b) The extracted insertion loss of the TMN. The inset shows the input impedance Z_{in}.

To achieve the required permittivity, and hence capacitance changes, the tuning voltages needed to be extended up to 400 V, which is a typical value for breakdown through air at short distances [Pas89]. To overcome this circumstance, the structure was coated with a SU8 polymer photo resist with a documented breakdown voltage of more than 400 V/µm [Mel+09]. Fig. 5.25a shows the microscope image of the processed structure.

For small signal characterization with GSG probes, the circuit provides contact pads with 1250 µm-pitch at the the in- and output. Further, a large contact pad at the input of the TMN was realized for the implementation with the transistor cells. The circuit also features a tapering region from 1250 µm-pitch GSG probes to 150 µm-pitch GSG probes at the output, which are used in the large signal measurement setup. To bias the varactors, pin probes were used to contact the pads U_{B_1} and U_{B_2}. External bias tees were used to provide reference bias potential to the signal (S) and ground (G).

The TMN was characterized with a VNA (Anritsu Vector Star 37397C) in a 50 Ω system impedance. However, the network is intrinsically designed to have an input impedance different from 50 Ω. Therefore, to estimate the insertion loss of the circuit, the S-parameters measured in a 50 Ω system impedance were renormalized in accordance to [Fri94]. The post-processed small signal measurement results are shown in Figure 5.25(b). The insertion loss (IL) varies in a range between 0.7 dB–1.1 dB depending on the frequency and tuning state of the varactors. Consequently, a loss induced PAE reduction between 15 % and 19 % can be expected for the transistor. The absolute change can be calculated in accordance to:

$$\Delta PAE = \left(1 - 10^{-IL/10}\right) \cdot PAE_0. \tag{5.17}$$

Here, PAE_0 is the efficiency achieved with lossless tuners. The tunability of the varactors allows the change in input impedance of the network, as shown in the inset of Figure 5.25(b). Any input impedance covered by the shaded areas can be reached by tuning the bias voltage in a range between 0 V and 400 V. The leakage current in static operation at maximum voltage does not exceed 1 µA. Hence the power consumption of the bias circuitry can be neglected.

5.4 Hybrid Implementation into a GaN HEMT

Implementation of the thick film TMN

To evaluate the performance of a tunable PA module, the tunable matching network was mounted on the same flange with the transistor and wire-bonded to three of the five transistor cells available on the transistor die, see Fig. 5.26. The flange itself was mounted in a test-fixture with a coaxial connection to the source-tuner and a bias-tee for the gate biasing. The source tuner was placed in a position where it provided good gain and stable operation of the transistor. At the load side, the TMN was wire-bonded to a stripline and connected to a mechanical load-tuner which allows corrections in fabrication tolerances of the TMN. The drain voltage was applied using an additional bias-tee. Figure 5.26 shows the schematic of the setup. The importance of the lowpass implementation of the tunable matching network becomes apparent at this point. The varactor biasing was provided through soldering fins that were bonded to the TMN with 5 mm long single bond wires that, due to their inductance, also should provide efficient RF isolation. The transistors were fed at 2 GHz and the input power was swept in a range from 21 dBm to 27 dBm. Both varactors were tuned in a range between 0 V and −400 V, while the output power was measured for each configuration independently. The polarity of the varactor bias voltage was reversed for implementation reasons. Due to symmetric behavior of BST, this does not influence the performance shown in Fig. 5.25.

Fig. 5.27(a) shows the results of PAE at saturated output power (P_{sat}) and in 6 dB power back-off measured at 15 V drain voltage at 2.0 GHz. The distinct points for maximum efficiency and corresponding output power (P_{out}) are pointed out at

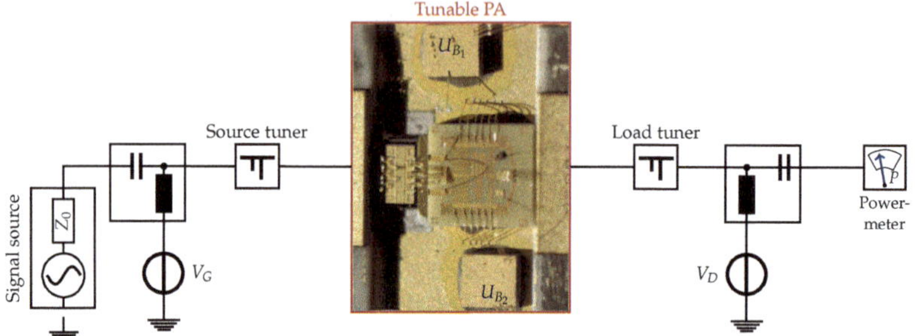

Figure 5.26: Measurement setup of the implemented tunable PA. Note that the drain voltage (15 V or 28 V) defines the reference potential for the bias at the signal electrode therefore, to fully exploit the tunability, the bias voltage for the matching network was performed with negative bias voltages.

saturation and back-off conditions to allow comparison of the performance. A maximum efficiency of $\eta = 47.1\,\%$ is measured at the varactor bias (280 V, 400 V). The efficiency drops to $\eta = 37.5\,\%$ for the same varactor bias when the input power is reduced by 6 dB. However, changing the tuning voltage to (400 V, 160 V) increases the efficiency to $\eta = 40.4\,\%$. This corresponds to an improvement of 2.9 %-points. in PAE. The same analysis is carried out for the output power in Fig. 5.27(b). Here, the output power in back-off condition can be improved by 0.3 dB. Generally, with this approach a PA designer can externally adjust the amplifier to operate at maximum efficiency or maximum output power either in saturation or back-off power. However, even with different load-tuner configurations the optimum impedance cannot be fully reached as indicated in the graphs (the optimum always lies at the corner of the plot). This implies that the bond wire inductance can be further adjusted. The main reason for the offset is assumed to come from the HEMT and bond wire tolerances. At the same time, an increased tunability of varactors would allow larger transformation ranges, and thus reduced sensitivity to fabrication tolerances.

The same measurements were repeated at 28 V and 40 V drain voltage. At 28 V, a maximum output power of 40.1 dBm with a peak efficiency of $\eta = 41\,\%$ was measured. The efficiency drop is assumed to be caused by further mismatch. At 40 V drain voltage a peak power of 41.6 dBm was measured, which demonstrates the high power capability of the TMN. However, at this power level the bias supply bond wires were destroyed, which is caused by insufficient RF- and DC decoupling.

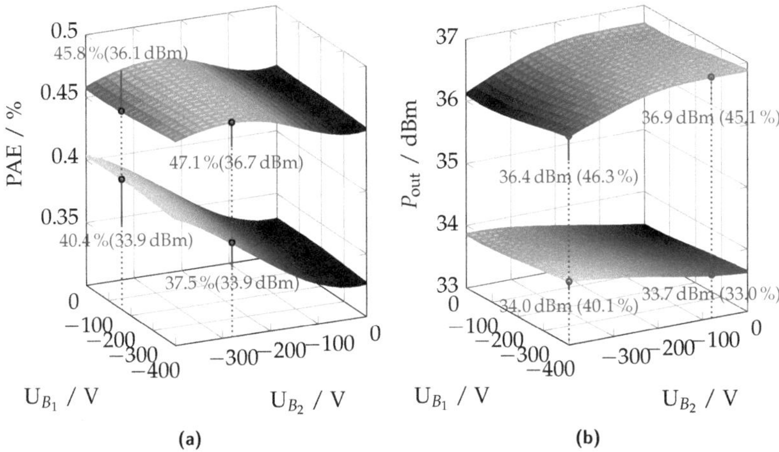

Figure 5.27: (a) Measured efficiency and (b) output power of the tunable transistor at saturation and 6 dB back-off power at 15 V drain voltage over the applied bias voltage of both varactors. The upper surface shows the tunability at approximately 3 dB compression. The lower surface shows the back-off efficiency response of the module.

A bond wire length of 5 mm can be approximated with 5 nH, which is at least factor 3 less than the investigations in Section 3.2.3 have suggested. This leakage of RF power into the biasing circuit, also reduces the maximum RF power and PAE.

The measurement setup utilized for the characterization was limited to 2 GHz, therefore, the response of frequency tunability has only been carried out as a harmonic balance simulation. Fig. 5.28 shows the estimated performance over frequency. Around 2 GHz no significant variation can be observed. The varactor bias voltage was adjusted to obtain maximum PAE and kept constant, while sweeping the frequency, see dashed lines in Fig. 5.28. Below 1.6 GHz and above 2.2 GHz improvements can be achieved with up to 2 dB in output power and around 8 %-points in PAE. The measured PAE at 2 GHz is slightly reduced compared to the simulation, while the measured P_{out} is slightly higher than predicted by simulation. This effect can be attributed to a mismatch of bond wire inductance in the simulation and the measurement. The results are still in good agreement, therefore it can be concluded that the simulated behavior qualitatively shows the nature of the module.

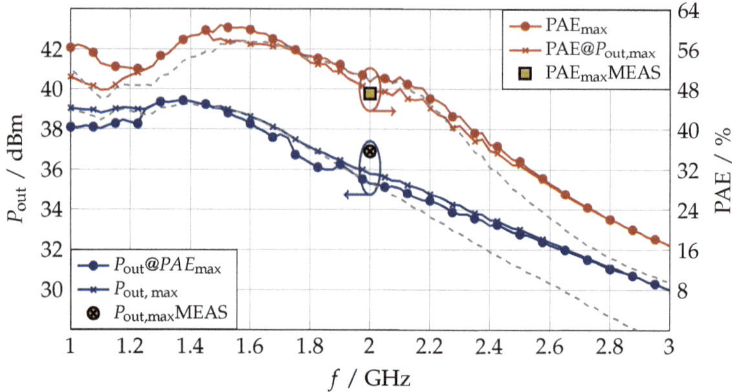

Figure 5.28: Estimated performance of the module versus frequency. The dashed line shows a case where the TMN was tuned to obtain maximum PAE at 2 GHz and kept constant, while sweeping the frequency.

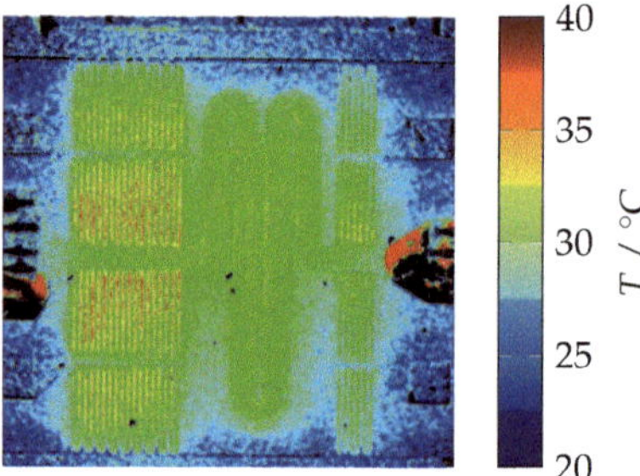

Figure 5.29: Thermal microscope image of the TMN exposed to 40 dBm RF power. The meander line significantly contributes to the losses of the network.

However, the improvements are still not sufficient to justify the additional effort and insertion loss of the TMN, especially with regard to the high varactor bias voltage. Assuming a $PAE_0 = 70\%$, in combination with the losses of a non-tunable matching network (0.5 dB) would lead to a PAE reduction of 7 %-points, see (5.17). The gain of the tunable matching network of about 3 % in PAE over 6 dB dynamic and an estimated average improvement over frequency of $\overline{\Delta\text{PAE}(f)} \approx 3\,\%$ does not justify the additional effort at the expense of additional PAE reduction (about $-6\,\%$-points). To obtain a competitive solution, the insertion loss has to be further reduced.

To investigate the source of losses, the network was driven under high RF power (40 dBm) and monitored with a thermal imaging microscope. Areas with increased temperature can be considered as a source of electromagnetic losses. The photograph, shown in Fig. 5.29, reveals that a lot of heat is dissipated in the region of IDC varactors. Here, the losses can only be reduced by material optimization. However, the meander line shows significant heating, which can be attributed to the fact that the meander itself is fabricated on top of a thin, lossy BST layer. Consequently, the insertion loss can be reduced by locally applying the BST layers to regions where they are required. A feasible technology to realize this, has been investigated in [Fri14] with locally ink-jet printed BST.

Implementation of the thin film TMN

The filter theory-based TMN, presented in Section 5.2.2, was mounted next to a 5-cell transistor bar, and a single cell was wire bonded to the TMN, see Fig. 5.30. Compared to Fig. 5.26, the measurement setup was modified to allow for GSG probe contact instead of coaxial connectors. This measurement setup was also calibrated over a broader frequency range to allow more accurate comparison between simulation and measurement. The transistor was biased through a bias-tee in Class-AB with a gate-source voltage $V_{GS} = -2.0\,\text{V}$ and a drain-source voltage $V_{DS} = 28\,\text{V}$. The measured quiescent current was approximately $I_{qs} \approx 50\,\text{mA}$. At the output, a load tuner is placed and set to $50\,\Omega$ impedance. The harmonics are left untreated with the tuner. A power-meter is used to detect the delivered power. A spectrum analyzer is applied to keep track of the linearity of the system. The tunable PA module was contacted with $150\,\mu\text{m}$-pitch GSG probes and measured at frequencies from 1 GHz to 2.5 GHz for different input power levels.

At 1 GHz the TMN is at its lower cutoff frequency, while at frequencies $f > 2.5\,\text{GHz}$ the insertion loss (see Fig. 5.16) is expected to dominate the modules performance. Therefore, the PA was measured only up to 2.5 GHz where IL $< 1\,\text{dB}$ is expected.

For each input power level and frequency, the output power P_{out} was measured for the bias voltages $U_{B_{1/2}} \in 0..20\ V$. The measurement was performed at distinct frequency points $f \in [1.0, 1.3, 1.56, 1.8, 2.0, 2.14, 2.5]$ GHz. Fig. 5.31 summarizes the results.

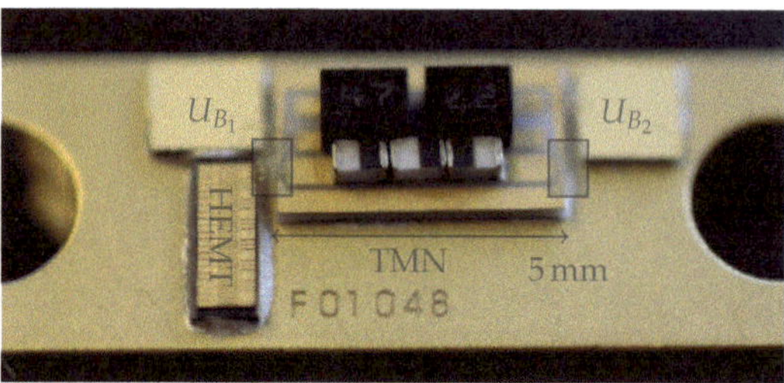

Figure 5.30: Tunable transistor module implementing a single GaN HEMT cell with a tunable matching network. The gray shaded area show the contact position for the GSG probes.

For each frequency point the tunable PA module was tuned to deliver maximum PAE with the corresponding output power P_{out} or maximum delivered output power $P_{out,max}$ with the corresponding PAE. At 1.8 GHz a maximum of $\eta_{max} = 65.7\%$ was measured, while the maximum power $P_{out,max} = 37.5$ dBm was observed at 1.56 GHz. At a single frequency point, an improvement of up to 7.2 %-points in PAE or 1 dB in P_{out} can be achieved by tuning the impedance network from PAE_{max} to $P_{out,\,max}$. For benchmark purposes, the graph includes the performance of a bare GaN HEMT cell, which was ideally matched with a load-tuner at 2 GHz, yielding maximum PAE $\approx 70\%$ or $P_{out} = 38$ dBm. Consequently, a reduction of 10 %-points in efficiency can be attributed to the losses in the TMN, see (5.17). The difference in output power of approximately 2 dB can be partly explained by the TMN losses (0.7 dB), which were extracted from a small signal on-wafer measurements of the TMN without the transistor. The remaining difference is assumed to be due to variations between the wafers used for measurement of the tunable module and the reference transistor, compare the harmonic balance simulation of the GaN HEMT cell in combination with the S-Parameter measurements in Fig. 5.20. Below 1.4 GHz the performance is strongly reducing, which is primarily caused by insufficient harmonic suppression, which has also been observed in the synthesis, see Fig. 5.16 and also in the harmonic balance simulation, see Fig. 5.20.

When varying frequency in the range from 1 GHz to 2.5 GHz, the PAE can be increased by up to 15 %-points and P_{out} by more than 2 dB. Fig. 5.32 summarizes the obtained results. It compares η_{max} and $P_{out,max}$ from Fig. 5.31 to a scenario where the TMN is tuned for a different operation frequency and depicts the difference, hence possible improvements. The dashed line in Fig. 5.32 is optimized for η_{max} at

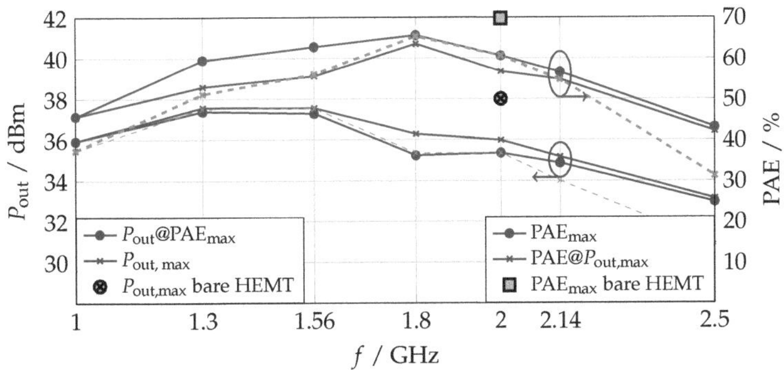

Figure 5.31: Measured maximum drain efficiency PAE and output power P_{out} achieved by electrically tuning of the matching network over the frequency range from 1 GHz to 2.5 GHz. For comparison, maximum PAE and P_{out} of a bare HEMT, ideally matched with lossless load tuners are shown. The dashed lines denote P_{out} and PAE versus frequency with TMN tuned for 2 GHz.

2 GHz, therefore at 2 GHz no further improvements are possible, as it already set to the maximum point. The bias voltage of the TMN is kept for this configuration and measuring at different frequencies leads to an efficiency drop (or possible improvements by tuning) of approximately 8 %-points at 2.5 GHz or at 1 GHz. Depending on the targeted operation frequency, the performance reduces even further and from all possible configurations the one with maximum possible improvement was picked. The same approach was followed for the improvements in delivered output power. An average improvement $\overline{\Delta PAE} = 10.2$ %-points and $\overline{\Delta P_{out}} = 0.9$ dB can be achieved throughout the measured frequency range.

Also, improvements can be achieved throughout the measured dynamic range, which is summarized in Fig. 5.33 for 2 GHz. As a reference, an ideally power matched bare GaN HEMT is shown, which delivers $P_{out} = 38$ dBm (corresponding PAE $= 62$ %). Here, the observed difference in output power can also only be explained by wafer variations. In total, an average improvement of $\overline{\Delta PAE} = 3.3$ %-points and $\overline{\Delta P_{out}} = 0.45$ dB.

The linearity was investigated in a CW two-tone measurement at 2 GHz with 1 MHz tone spacing. Fig. 5.34 shows the measured fundamental power and the intermodulation distortion product of third order (IM3). An output third order intercept

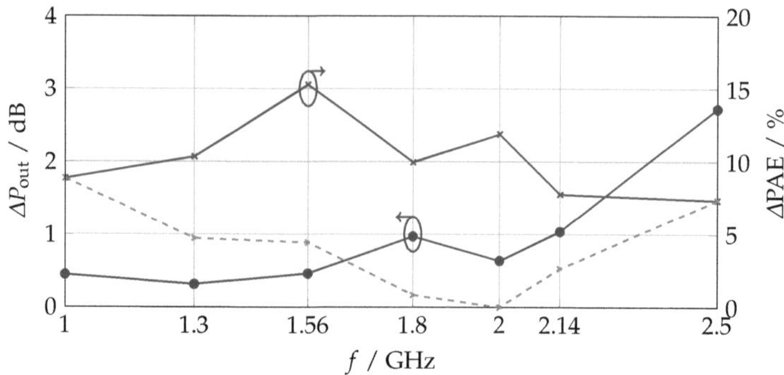

Figure 5.32: Measured improvements in PAE and P_{out} of the tunable PA versus frequency f. At each f, the maximum achievable value is compared to the case where the TMN is tuned to a different operation frequency. Dashed line shows the possible improvements in PAE over f if the TMN is tuned for 2 GHz.

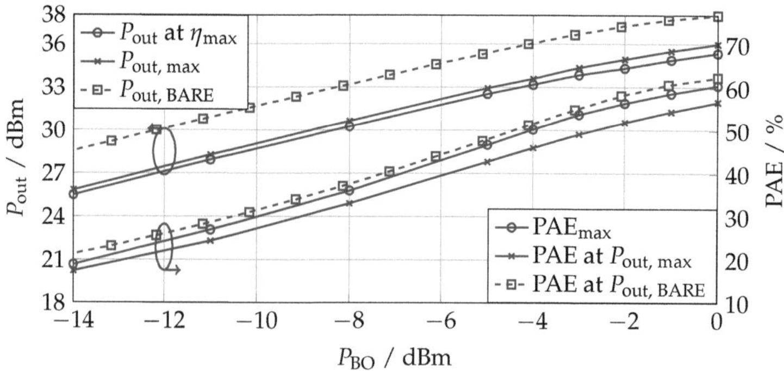

Figure 5.33: Performance of the tunable PA module over the measured dynamic range at 2 GHz. An ideally loss-less matched bare HEMT device is used as a reference.

point (OIP3) at 39 dBm is obtained by extrapolation. The value is dominated by the non-linearity of the transistor which is typically around $\text{IP3}_{\text{HEMT}} \approx 36..39\,\text{dBm}$.

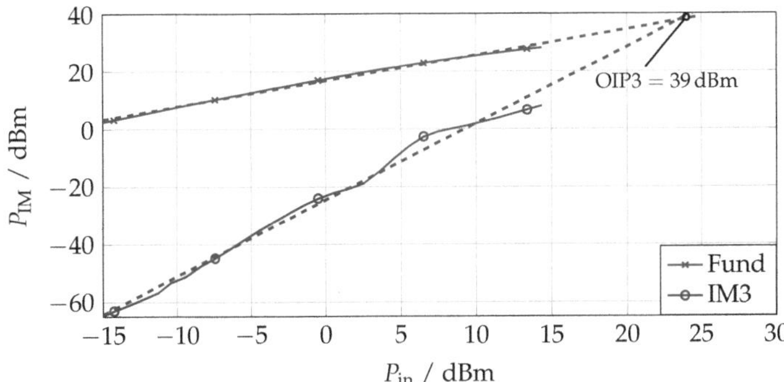

Figure 5.34: Measured fundamental and IM3 power with the third order intercept point OIP3 = 39 dBm.

The achieved performance improvements of the tunable matching networks, utilizing thin film varactors allow PAE improvements, that for the first time justify the additional effort. Especially over a wide frequency range significant improvements can be achieved. However, the results not necessarily rely on the synthesis method that was used for the thin film network. The proposed CAD-aided network synthesis suffers from the fact that the IDC varactors are inferior in both, quality factor and tunability. The implemented inductances are also inferior to the lumped components. Consequently, it is difficult to make conclusive statements on the quality of the two different synthesis approach. Both require extensive numerical optimization. The filter approach is currently only limited to synthesis, where only frequency dependence can be considered during the design. To include a second impedance dependence requires numerical optimization. The CAD-aided approach allows direct optimization, which is strongly depending on the formulation of the cost function. However, since this method can consider parasitics as well as several optimization dimensions, it is expected to give better results compared to the filter-based synthesis. This will be investigated in future works.

6 Conclusion and Outlook

A crucial goal for any component under high power operation remains the reduction of dielectric losses. Beside the undesired reduction of transmission, when implementing lossy elements, high power operation can ultimately lead to thermal destruction. Within the scope this work, novel composite material, combining low-Q but tunable barium strontium titanate with untunable high-Q magnesium boron oxide has been investigated. The goal is the reduction of the dielectric losses, while maintaining the electric tunability of the mixture. Sophisticated, CAD-aided physical model has been developed to describe the dielectric properties of the compound. The proposed model introduces an additional degree of freedom and allows statements on the statistics of the material over a wide range of mixture compositions. Further, the novel model has been compared to other analytical models, commonly used to describe the dielectric properties of mixtures, showing the benefits of the new approach. Comparison between simulation and measurements show good agreements, which previous models are unable to achieve.

To allow the implementation of tunable components, based on ferroelectric material, advanced numerical models have been implemented and allow accurate prediction on tunability on a physical level. Properties such as the geometric tuning efficiency [Mau+11], which have been previously investigated by time consuming inhouse 2D finite elements time domain methods have been verified and replaced with macro-based 3D fullwave analysis in CST Studio Suite (CST). Further, a numerical model utilizing the method of moments has been implemented in Agilent ADS. An approximation of the tunability has been implemented and allows the simulation of planar and multi layer components, that can be easily deployed in a circuit simulator and allow accurate prediction of tunable circuit behavior. The impact of resistive and inductive varactor biasing has been investigated and guidelines for cascading varactors are suggested.

Further, first systematic investigations of tunable ferroelectric varactors in parallel-plate configuration (MIM) and inter digital capacitors (IDCs) were performed for high power applications. The thermal conductivity of porous ceramic BST films has been determined for the first time with a thermal conductivity of $0.5\,\mathrm{W}/(\mathrm{m\,K})$, which is orders of magnitude below the previously expected value [Mau11]. Extensive thermal analysis, considering the obtained value, has been carried out for IDC varactors and MIM varactors. The results, obtained by thermal co-simulation in CST predict a loss-induced temperature increase of $10\,°C$ at an RF power level of $0.5\,W$, which is factor 100 lower than predicted in previous investigations [Mau11]. This result has been supported with measurements. The multi layer architecture shows an intrinsically better performance, as the RF power, and hence the heating, is not concentrating only between the IDC fingers. A rated power handling capability of $100\,W/mm$ has been predicted by simulations and verified with large signal measurements. The results are put in context to novel commercial thin film BST varactors, showing large potential for the thick film technology.

The linearity of the tunable IDC and MIM varactors was investigated with harmonic balance simulation in combination with polynomial models of the tuning behavior. The simulations predict an OIP3 $\geq 59\,dBm$ for a 1 pF IDC varactors with a typical finger gap of $15\,\mu m$. The reduced electrode gap of MIM varactors (approximately $4\,\mu m$) yields a reduced OIP3 $\geq 55\,dBm$ for a single varactor with a normalized capacitance of 1 pF. Even further reduced BST layer thickness of the commercial multi layer thin film varactors (STMicroelectronics) has an even more drastic reduction of OIP3 to 26 dBm for a single normalized varactor of 1 pF. The influence of cascading the varactors has been investigated with the goal to increase the power handling capability as well as to increase the OIP3. Measurements conducted around $2\,GHz$ have shown an $OIP3_{IDC} = 88\,dBm$, $OIP3_{MIM} = 74\,dBm$ and $OIP3_{STM} = 64\,dBm$. Harmonic balance simulations were performed to support these high values.

In conclusion, the viability study of BST components suggested a direct application in a matching network for high power operation. The elements were implemented as part of tunable impedance matching network, that was compact enough to fit inside a transistor package. For the design of the tunable impedance matching network, two routines were followed. A filter synthesis based approach, which has been extended to include tunable elements and a purely numerical optimization approach, utilizing advanced algorithms to synthesize a network that comes closest to a predefined characteristics in the impedance space. The former approach requires elements with reduced parasitic effects and has been therefore implemented in MIM technology. However, the fabrication tolerances of the screen printing technology did not allow accurate realization of required capacitance values, which had the effect of deteriorated RF performance. Further, the delamination of the layers did not allow the integration of the network with the transistor cell. Alternatively, the circuit has been realized with commercial thin film multi layer varactors. The realized circuit shows an insertion loss around $0.5\,dB$ up to $1.6\,GHz$ and below $1\,dB$ up to $2.5\,GHz$. The realized tunable PA module, implementing a tunable impedance matching network with a single GaN HEMT cell showed excellent performance with power added efficiency (PAE) up to 65 % around 1.8 GHz and PAE improvements over the frequency span between 1 GHz and 2.5 GHz of up to 15 %-points. However, this synthesis method, intrinsically, does not allow to explicitly address the input power dependent impedance variations. This issue has been addressed with the impedance space synthesis method, which has been implemented in thick film technology using IDC varactors. Building blocks of circuit elements have been designed and numerically combined to match three of the GaN HEMT cells directly to 50 Ω. The synthesized circuit shows an insertion loss below 1 dB around 2 GHz, which is slightly higher than the thin film implementation and can be attributed to slightly higher losses of the thick film BST layer. However, the networks synthesized with the accurate models, designed in this work, allowed a tremendous reduction of insertion losses of a matching network from approximately 9 dB [Mau11] down to below 1 dB. The fabricated network was wire bonded to three GaN HEMT cells, yielding PAE of up to 47 % and a maximum output power of close to 37 dBm at 2 GHz. At 6 dB power back-off level, an improvement of PAE up to 3 % or up to 0.3 dB in output power could be achieved by tuning the varactors. Output power of more than 41.5 dBm was measured, showing the high power capability of this technology. Using the sophisticated models, developed within this work, tunable impedance matching networks with very low overall insertion losses, small enough to fit inside a transistor package have been demonstrated. The additional effort in manufacturing tunable transistors is justified due to the excellent power added efficiency values.

The models, developed within this work, are powerful tools to engineer novel tailored materials and to accurately design and simulate functional material-based circuits and systems. It has been shown, that the continuously tunable BST varactors can improve the overall efficiency and output power of transistors and sustain large RF power level, while maintaining the linearity requirements of RF communication systems. Tunable circuits for the input of the transistor cell can be realized and a fully tunable and fully matched, efficient transistor is feasible with the proposed models. Also, other application fields, where high power operation is sometimes required, e.g. tunable filter, phase shifters or antenna matching networks of base stations can significantly benefit from the results obtained in this work.

A Mathematical Appendix

Effective dielectric properties of layered composites

Employing the continuity of electrical displacement $D = (\varepsilon' - j\varepsilon'')E$ with $D_{\perp,1} \overset{!}{=} D_{\perp,2}$ in combination with (3.4) results in:

$$\varepsilon'_{\text{eff}}(E,q) \left| \frac{D}{\varepsilon'_{\text{eff}} - j\varepsilon''_{\text{eff}}} \right|^2 = q \cdot \varepsilon'_f(E_f) \left| \frac{D}{\varepsilon'_f - j\varepsilon''_f} \right|^2 + (1-q) \cdot \varepsilon'_d \left| \frac{D}{\varepsilon'_d - j\varepsilon''_d} \right|^2$$

$$\varepsilon''_{\text{eff}}(E,q) \left| \frac{D}{\varepsilon'_{\text{eff}} - j\varepsilon''_{\text{eff}}} \right|^2 = q \cdot \varepsilon''_f(E_f) \left| \frac{D}{\varepsilon'_f - j\varepsilon''_f} \right|^2 + (1-q) \cdot \varepsilon''_d \left| \frac{D}{\varepsilon'_d - j\varepsilon''_d} \right|^2 ,$$

$$\text{(A.1)}$$

which can be further simplified to:

$$\frac{\varepsilon'_{\text{eff}}(E,q)}{\varepsilon'_{\text{eff}}(E,q)^2 + \varepsilon''_{\text{eff}}(E,q)^2} = \frac{q \cdot \varepsilon'_f(E_f)}{\varepsilon'_f(E_f)^2 + \varepsilon''_f(E_f)^2} + \frac{(1-q) \cdot \varepsilon'_d}{\varepsilon'^2_d + \varepsilon''^2_d} \tag{A.2}$$

$$\frac{\varepsilon''_{\text{eff}}(E,q)}{\varepsilon'_{\text{eff}}(E,q)^2 + \varepsilon''_{\text{eff}}(E,q)^2} = \frac{q \cdot \varepsilon''_f(E_f)}{\varepsilon'_f(E_f)^2 + \varepsilon''_f(E_f)^2} + \frac{(1-q) \cdot \varepsilon''_d}{\varepsilon'^2_d + \varepsilon''^2_d}. \tag{A.3}$$

To decouple this set of equations one now can divide (A.3) over (A.2) and using the relation $\tan \delta = \varepsilon'' / \varepsilon'$ leads to:

$$\tan \delta_{\text{eff}}(E,q) = \left[\frac{q \cdot \varepsilon''_f(E_f)}{\varepsilon'_f(E_f)^2 + \varepsilon''_f(E_f)^2} + \frac{(1-q) \cdot \varepsilon''_d}{\varepsilon'^2_d + \varepsilon''^2_d} \right]$$

$$\times \left[\frac{q \cdot \varepsilon'_f(E_f)}{\varepsilon'_f(E_f)^2 + \varepsilon''_f(E_f)^2} + \frac{(1-q) \cdot \varepsilon'_d}{\varepsilon'^2_d + \varepsilon''^2_d} \right]^{-1}. \tag{A.4}$$

Rewirting (A.4) leads to:

$$\tan \delta_{\text{eff}}(E,q) = \frac{q\varepsilon''_f(E_f) \cdot (\varepsilon'^2_d + \varepsilon''^2_d) + (1-q)\varepsilon''_d \cdot \left[\varepsilon'_f(E_f) + \varepsilon''_f(E_f) \right]}{q\varepsilon'_f(E_f) \cdot (\varepsilon'^2_d + \varepsilon''^2_d) + (1-q)\varepsilon'_d \cdot \left[\varepsilon'_f(E_f)^2 + \varepsilon''_f(E_f)^2 \right]}, \tag{A.5}$$

and using the relation $\varepsilon'' = \tan\delta\varepsilon'$ (A.5) reads:

$$\tan\delta_{\text{eff}}(E,q) = \frac{q\varepsilon'_d \tan\delta_f(E_f) \cdot (1 + \tan\delta_d^2) + (1-q)\tan\delta_d\varepsilon'_f(E_f) \cdot \left[1 + \tan\delta_f^2(E_f)\right]}{q\varepsilon'_d \cdot (1 + \tan\delta_d^2) + (1-q)\varepsilon'_f(E_f) \cdot \left[1 + \tan\delta_f(E_f)\right]}.$$
$$(A.6)$$

For the case of moderate losses $\tan\delta < 0.1 \rightarrow \tan\delta^2 \ll 1$ (A.6) simplifies to the equation given in a wide range of publications (e.g. [Ast+03; Tag+03; She+06]). It has to be noted that the equation derived here, is the generalization of the equation given in literature:

$$\tan\delta_{\text{eff}}(E,q) = \frac{q\varepsilon'_d \tan\delta_f(E_f) + (1-q)\tan\delta_d\varepsilon'_f(E_f)}{q\varepsilon'_d + (1-q)\varepsilon'_f(E_f)}.$$
$$(A.7)$$

To derive the exact behavior of the effective permittivity in layered composites (A.2) can once again be rewritten using the relation $\varepsilon'' = \tan\delta\varepsilon'$ leading to:

$$\frac{1}{\varepsilon'_{\text{eff}}(E,q) \cdot (1 + \tan\delta_{\text{eff}}^2(E,q))} = \frac{q}{\varepsilon'_f(E_f) \cdot (1 + \tan\delta_f^2(E_f))} + \frac{(1-q)}{\varepsilon'_d \cdot (1 + \tan\delta_d^2)}. \quad (A.8)$$

Using the same condition for $\tan\delta < 0.1 \rightarrow \tan\delta^2 \ll 1$ leads to the equation given in the literature:

$$\frac{1}{\varepsilon'_{\text{eff}}(E,q)} = \frac{q}{\varepsilon'_f(E_f)} + \frac{(1-q)}{\varepsilon'_d}. \quad (A.9)$$

Again it has to be noted that the (A.8) is the generalization of the formula given in literature (e.g. [She+06]). Combining (A.8) and (A.6) the exact solution for $\tau_{\text{eff}}(E)$ can be calculated. For simplicity reasons, the tunability of layered composite is calculated with the approximation (A.9) since the expected loss of the ferroelectric and the dielectric material is in the range of $\tan\delta < 0.01$.

$$\begin{aligned}
\tau_{\text{eff}}(E,q) &= \frac{(\varepsilon'_f(0) - \varepsilon'_f(E_f)) \cdot q \cdot \varepsilon'_d}{\varepsilon'_f(0) \cdot ((1-q) \cdot \varepsilon'_f(E_f) + q \cdot \varepsilon'_d)} \\
&= \tau_f(E_f) \cdot \frac{q \cdot \varepsilon'_d}{(1-q) \cdot \varepsilon'_f(E_f) + q \cdot \varepsilon'_d}.
\end{aligned}$$
$$(A.10)$$

Effective loss with *spherical inclusions*

The goal is to solve the $\nabla^2\varphi_0^m = 0$ in V_m, which is the volume of the host medium and $\nabla^2\varphi_0^i = 0$ in V_i which is the volume of the spherical inclusion. The solution in spherical coordinate system is well known in classical field theory [Smy68]:

$$\varphi_0^m = -E_0(r - br^{-2})\cos\vartheta$$
$$\varphi_0^i = -cE_0 r\cos\vartheta,$$

(A.11)

with $b = (\varepsilon_i' - \varepsilon_m')\rho^3/(\varepsilon_i' + 2\varepsilon_m')$ and $c = 3\varepsilon_m'/(\varepsilon_i' + 2\varepsilon_m')$. Further, E_0 is the externally applied field and ρ the radius of the spherical inclusion. The scalar potential φ which leads to an E-field $\mathbf{E} = \nabla\varphi$.

In accordance to (3.2) and (3.1) in spherical coordinate system follows:

$$\varepsilon_{\text{eff}}'' V E_0^2 = \int_\pi^0 2\pi\sin\vartheta d\vartheta \left[\underbrace{\varepsilon_i'' \int_0^\rho r^2 \left|\nabla\varphi_0^i\right|^2 dr}_{I_1} + \underbrace{\varepsilon_m'' \int_\rho^R r^2 \left|\nabla\varphi_0^m\right|^2 dr}_{I_2} \right].$$

(A.12)

The integration is separated into two parts: spherical inclusion I_1 from 0 to ρ and remaining host medium I_2 from ρ to R. For the sake of simplicity, each intergral is evaluated separately.

$$I_1 = \varepsilon_i'' \int_0^\rho r^2 \left|\nabla\varphi_0^i\right|^2 dr = \frac{1}{3}\varepsilon_i'' c^2 E_0^2,$$

$$I_2 = \varepsilon_m'' \int_\rho^R r^2 \left|\nabla\varphi_0^m\right|^2 dr = \frac{1}{6}E_0^2\varepsilon_m'' \left[\frac{(R^3 - \rho^3)(5b^2 + 2R^3\rho^3 + 3b^2\cos 2\vartheta)}{R^3\rho^3} \right.$$

(A.13)

$$\left. + 6b(1 + 3\cos 2\vartheta)\log\left(\frac{R}{\rho}\right) \right].$$

The full integral of (A.12) in all orientation yields:

$$\varepsilon_{\text{eff}}'' V E_0^2 = \int_\pi^0 2\pi\sin\vartheta\left[I_1 + I_2\right] d\vartheta$$

$$= E_0^2 \left[\underbrace{\frac{4}{3}\pi R^3 \varepsilon_m''}_{V} + 2\underbrace{\frac{4}{3}\pi b^2 \varepsilon_m''}_{V_i b'^2 \rho^3}\left(\frac{R^3 - \rho^3}{R^3\rho^3}\right) + \underbrace{\frac{4}{3}\pi\rho^3}_{V_i}\left(c^2\varepsilon_i'' - \varepsilon_m''\right) \right].$$

(A.14)

Using $b = b'\rho^3$, $V_i = 4/3\pi\rho^3$, $V = 4/3\pi R^3$ and defining volume fraction coefficient $q = V_i/V$ in accordance to Section 3.1.1 simplifies (A.14):

$$\varepsilon_{\text{eff}}'' = \varepsilon_m'' + 2q\varepsilon_m'' b'^2 \underbrace{\left(\frac{R^3 - \rho^3}{R^3}\right)}_{\approx 1} + q\left(c^2\varepsilon_i'' - \varepsilon_m''\right).$$

(A.15)

For small concentration of spherical inclusions the difference in radii is negligible and (A.16) further simplifies into:

$$\varepsilon''_{\text{eff}} = \varepsilon''_m + 2q\varepsilon''_m b'^2 + q\left(c^2\varepsilon''_i - \varepsilon''_m\right). \tag{A.16}$$

However, there are two types of corrections due to large, but limited volume V – small corrections of order R^{-3} and large corrections of order unity which appear near the surface [BB89].

$$\varepsilon''_{\text{eff}} = \varepsilon''_m + 2q\varepsilon''_m b'^2 + q\left(c^2\varepsilon''_i - \varepsilon''_m\right) + \underbrace{2q\varepsilon''_m b'}_{\text{Correction}}. \tag{A.17}$$

Note that due to symmetry between ε' and ε'' (A.17) yields:

$$\varepsilon'_{\text{eff}} = \varepsilon'_m = \frac{3q\varepsilon'_m\left(\varepsilon'_i - \varepsilon'_m\right)}{\varepsilon'_i + 2\varepsilon'_m}, \tag{A.18}$$

which equals Eq. 3.16 in Section 3.1.1. However, rewriting (A.17) using $\tan\delta = \varepsilon''/\varepsilon'$ gives:

$$\tan\delta_{\text{eff}} = \tan\delta_m + \frac{9q\varepsilon'_i\varepsilon'_m\left(\tan\delta_i - \tan\delta_m\right)}{\left(\varepsilon'_i + 2\varepsilon'_m\right)\left(\varepsilon'_i + 2\varepsilon'_m + 3q\left(\varepsilon'_i - \varepsilon'_m\right)\right)} \tag{A.19}$$

B Characterization Appendix

B.0.1 Thin Film Varactor Characterization

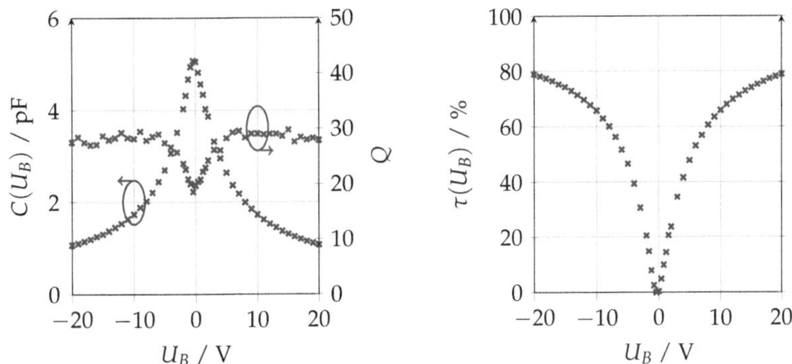

Figure B.1: (a) Capacitance and Q of a 3.3 pF commercial thin film varactor versus bias voltage U_B. (b) Tunability τ versus bias voltage. Values extracted at 2 GHz.

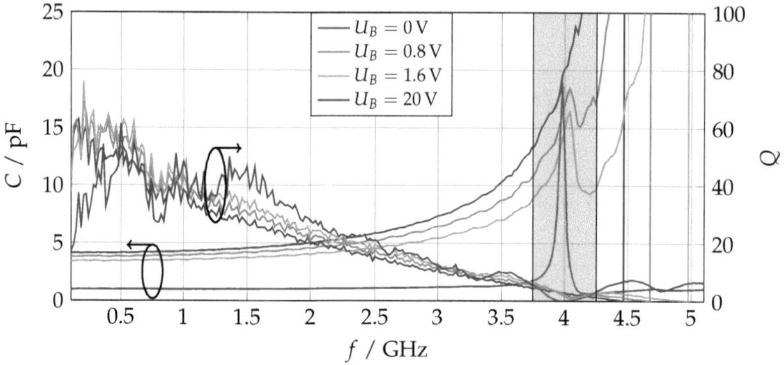

Figure B.2: Capacitance C and quality factor Q of a 3.3 pF commercial thin film varactor versus frequency f for different bias voltages. Note the acoustic resonance, that appears around 4 GHz when bias voltage is applied.

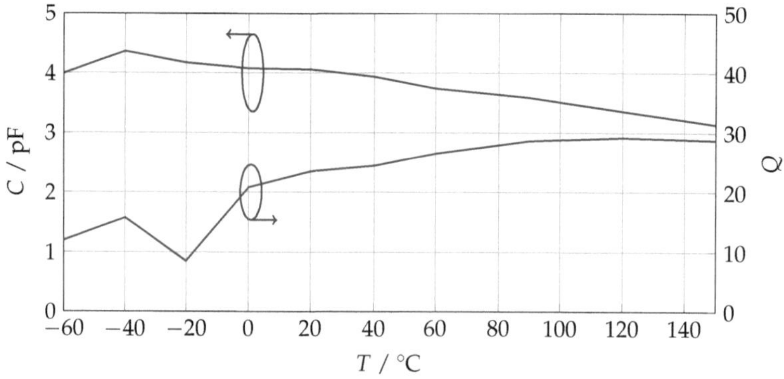

Figure B.3: Capacitance C and quality factor Q of a 3.3 pF commercial thin film varactor versus temperature T at 2 GHz and pre-biased with 2 V.

B.0.2 Thick Film Varactor Characterization

Uniplanar Varactor

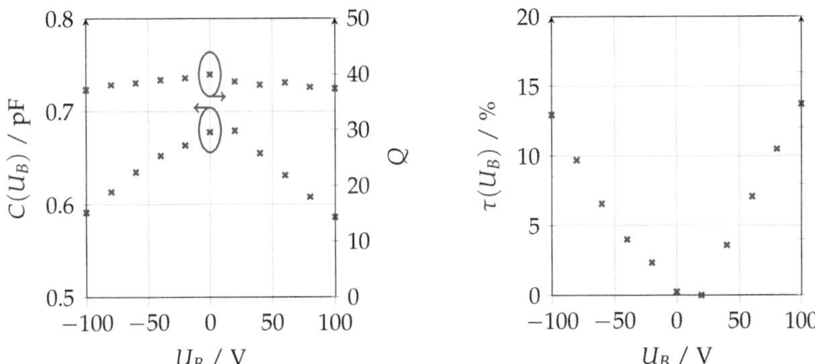

Figure B.4: (a) Capacitance and Q of a $\approx 0.7\,\mathrm{pF}$ IDC varactor on an $4\,\mu\mathrm{m}$ high FeF codoped BST layer versus bias voltage U_B. (b) Tunability τ versus bias voltage. Values extracted at $2\,\mathrm{GHz}$.

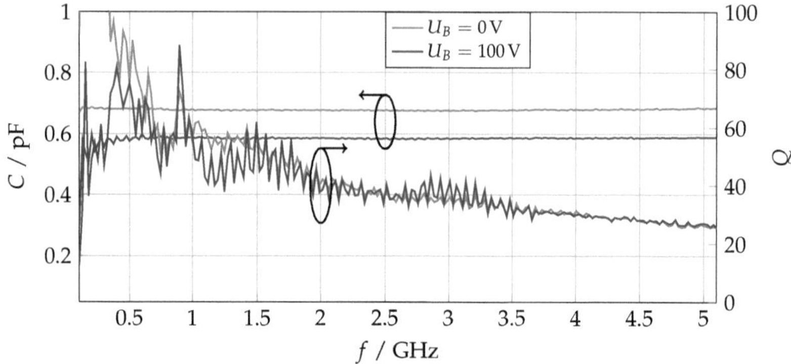

Figure B.5: Capacitance C and quality factor Q of a $\approx 0.7\,\mathrm{pF}$ IDC varactor on an $4\,\mu\mathrm{m}$ high FeF codoped BST layer versus frequency f for different bias voltages. Note that no acoustic resonances appear throughout the entire frequency range.

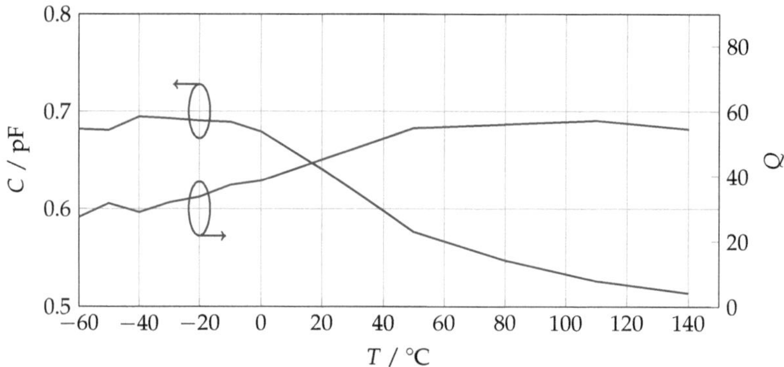

Figure B.6: Capacitance C and quality factor Q of a $\approx 0.7\,\mathrm{pF}$ IDC varactor on an $4\,\mu\mathrm{m}$ high FeF codoped BST layer versus temperature T at $2\,\mathrm{GHz}$ and pre-biased with $20\,\mathrm{V}$.

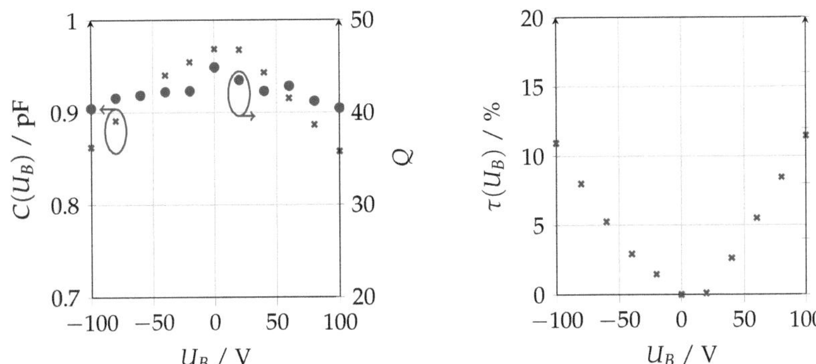

Figure B.7: (a) Capacitance and Q of a $\approx 1\,\text{pF}$ IDC varactor on an $8\,\mu\text{m}$ high FeF codoped BST layer versus bias voltage U_B. (b) Tunability τ versus bias voltage. Values extracted at $2\,\text{GHz}$.

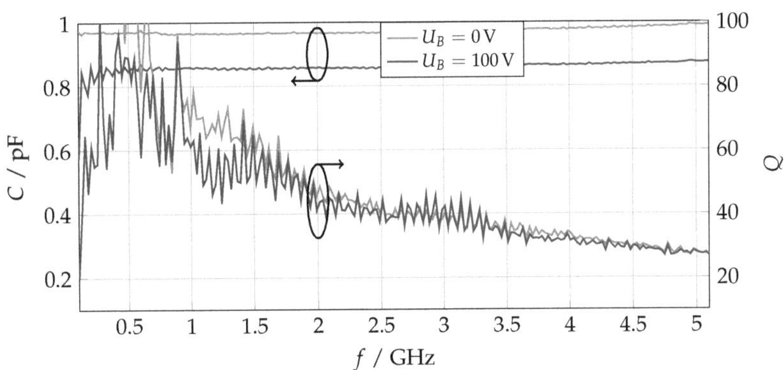

Figure B.8: Capacitance C and quality factor Q of a $\approx 1\,\text{pF}$ IDC varactor on an $8\,\mu\text{m}$ high FeF codoped BST layer versus frequency f for different bias voltages. Note that no acoustic resonances appear throughout the entire frequency range.

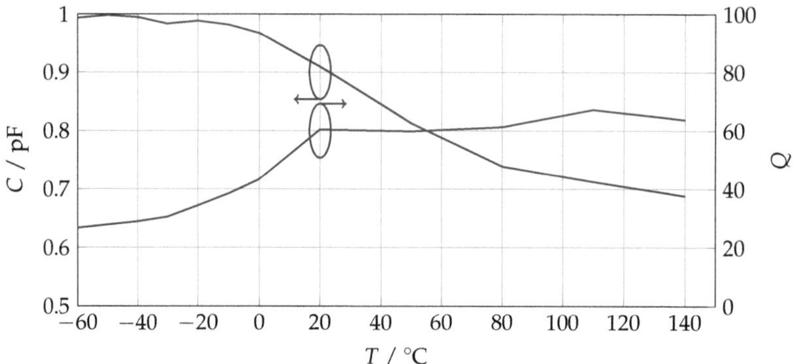

Figure B.9: Capacitance C and quality factor Q of a $\approx 1\,\mathrm{pF}$ IDC varactor on an $8\,\mu\mathrm{m}$ high FeF codoped BST layer versus temperature T at $2\,\mathrm{GHz}$ and pre-biased with $20\,\mathrm{V}$.

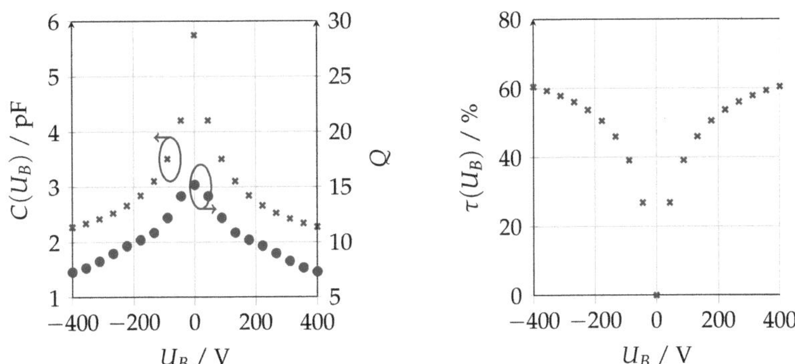

Figure B.10: (a) Capacitance and Q of a $\approx 5.5\,\mathrm{pF}$ MIM varactor on an $4\,\mu\mathrm{m}$ high CuF codoped, low temperature sintered BST layer versus bias voltage U_B. (b) Tunability τ versus bias voltage. Values extracted at $2\,\mathrm{GHz}$.

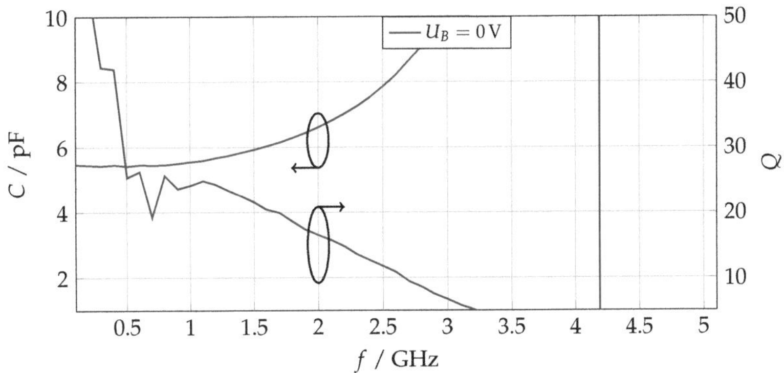

Figure B.11: Capacitance C and quality factor Q of a $\approx 5.5\,\mathrm{pF}$ MIM varactor with a $4\,\mu\mathrm{m}$ high CuF codoped, low temperature sintered BST layer versus frequency f. Note that no acoustic resonances appear throughout the entire frequency range.

B.0.3 Bulk Varactor Characterization

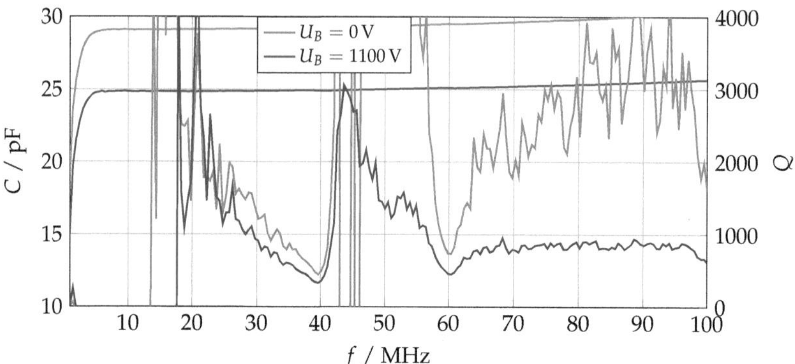

Figure B.12: Capacitance C and quality factor Q of a 28 pF bulk MIM varactor with a 0.5 mm thick FeF codoped BST/MgBO composite disk versus frequency f.

C Figure Appendix

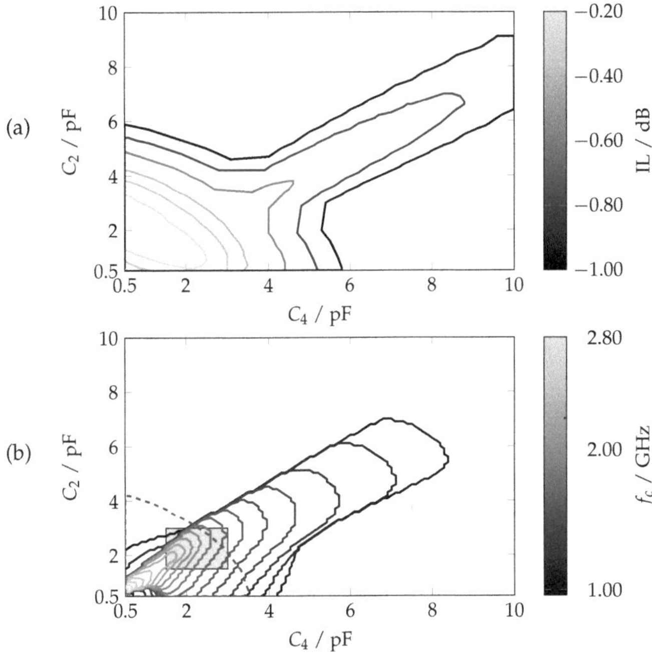

Figure C.1: Contour plot of the insertion loss (a) and cutoff frequency (b) versus the change in capacitance values C_2 and C_4 assuming lossy elements. The dashed lines in (b) give the possible tuning under the condition that the insertion loss is below 0.4 dB.

D Table Appendix

Table D.1: Table of element values for a Chebyshev filter with different ripple levels L_r and element numbers N.

N	g_1	g_2	g_3	g_4	g_5	g_6	g_7	g_8	g_9	g_{10}	g_{11}
					0.01 dB						
1	0.0960	1.0000									
2	0.4489	0.4078	1.1008								
3	0.6292	0.9703	0.6292	1.0000							
4	0.7129	1.2004	1.3213	0.6476	1.1008						
5	0.7563	1.3049	1.5773	1.3049	0.7563	1.0000					
6	0.7814	1.3600	1.6897	1.5350	1.4970	0.7098	1.1008				
7	0.7970	1.3924	1.7481	1.6331	1.7481	1.3924	0.7970	1.0000			
8	0.8073	1.4131	1.7825	1.6833	1.8529	1.6193	1.5555	0.7334	1.1008		
9	0.8145	1.4271	1.8044	1.7125	1.9058	1.7125	1.8044	1.4271	0.8145	1.0000	
10	0.8197	1.4370	1.8193	1.7311	1.9362	1.7590	1.9055	1.6528	1.5817	0.7446	1.1008
					0.3 dB						
1	0.5349	1.0000									
2	1.1805	0.6957	1.6967								
3	1.3713	1.1378	1.3713	1.0000							
4	1.4457	1.2537	2.1272	0.8521	1.6967						
5	1.4817	1.2992	2.3095	1.2992	1.4817	1.0000					
6	1.5016	1.3218	2.3790	1.4021	2.2427	0.8850	1.6967				
7	1.5138	1.3346	2.4131	1.4403	2.4131	1.3346	1.5138	1.0000			
8	1.5217	1.3427	2.4325	1.4587	2.4751	1.4336	2.2782	0.8969	1.6967		
9	1.5272	1.3481	2.4447	1.4691	2.5045	1.4691	2.4447	1.3481	1.5272	1.0000	
10	1.5311	1.3518	2.4529	1.4756	2.5210	1.4858	2.5037	1.4457	2.2937	0.9024	1.6967
					0.4 dB						
1	0.6213	1.0000									
2	1.2989	0.7046	1.8435								
3	1.4909	1.1180	1.4909	1.0000							
4	1.5650	1.2225	2.2537	0.8489	1.8435						
5	1.6006	1.2632	2.4315	1.2632	1.6006	1.0000					
6	1.6203	1.2833	2.4986	1.3553	2.3658	0.8789	1.8435				
7	1.6323	1.2947	2.5314	1.3892	2.5314	1.2947	1.6323	1.0000			
8	1.6402	1.3019	2.5500	1.4055	2.5910	1.3832	2.4000	0.8897	1.8435		
9	1.6456	1.3066	2.5617	1.4147	2.6193	1.4147	2.5617	1.3066	1.6456	1.0000	
10	1.6495	1.3100	2.5696	1.4204	2.6351	1.4294	2.6185	1.3938	2.4150	0.8947	1.8435
					0.5 dB						
1	0.6987	1.0000									
2	1.4029	0.7071	1.9841								
3	1.5963	1.0967	1.5963	1.0000							
4	1.6704	1.1925	2.3662	0.8419	1.9841						
5	1.7058	1.2296	2.5409	1.2296	1.7058	1.0000					
6	1.7254	1.2479	2.6064	1.3136	2.4759	0.8696	1.9841				
7	1.7373	1.2582	2.6383	1.3443	2.6383	1.2582	1.7373	1.0000			
8	1.7451	1.2647	2.6565	1.3590	2.6965	1.3389	2.5093	0.8795	1.9841		
9	1.7505	1.2690	2.6678	1.3673	2.7240	1.3673	2.6678	1.2690	1.7505	1.0000	
10	1.7543	1.2721	2.6755	1.3725	2.7393	1.3806	2.7232	1.3484	2.5239	0.8842	1.9841

E Lithography Appendix

E.1 Photo lithography

Pos. photo resist – AZ4562 for metal layers up to $3\,\mu m$ to $5\,\mu m$

Pre exposure bake – $90\,°C$ for $1\,min/\mu m$

UV exposure – approx. $30\,s$

Developer – AZ400 : H_2O (1:4) for $60\,s$ at room temperature ($23\,°C$)

Hard bake – $120\,°C$ for $15\,min$

Neg. photo resist – Ma-N 490 for metal layers up to $3\,\mu m$ to $5\,\mu m$

Pre exposure bake – $105\,°C$ for $1.5\,min/\mu m$

UV exposure – approx. $180\,s$

Developer – ma-D 332/s: for $300\,s$ at room temperature ($23\,°C$)

UV flood exposure – approx. $300\,s$

O_2-plasma purge to remove resisudes of photo resist and developer.

RF power – $200\,W$

O_2-gas flow – $100\,sccm$ at $20\,kPa$

duration – $30\,s$

E.2 Electroplating

Gold

Galvanic – Doduco Puramet 402

DC Current – $1\,\mathrm{mA/cm^2}$

Temperature – $35\,^\circ\mathrm{C}$

E.3 Etchants

Gold

Solution – I_2: KI : H_2O (1 g : 4 g : 40 ml)

Temperature – room temperature ($23\,^\circ\mathrm{C}$)

Chromium/Nickel

Solution – $(NH_4)_2\,[Ce\,(NO_3)_6]$: HNO_3: H_2O (82.3 g : 45 ml : 500 ml)

Temperature – room temperature ($23\,^\circ\mathrm{C}$)

F Measurement equipment

The following measurement equipment has been utilized within this work:

Type: Manufacturer, Resolution
Sourcemeter: Keithley 2612   max. output voltage: $\pm 200\,\mathrm{V}$, current resolution: $2\,\mathrm{pA}$
Picoammeter: Keithley 487   max. output voltage: $\pm 500\,\mathrm{V}$, current resolution: $10\,\mathrm{fA}$
Picoammeter: Keithley 2410   max. output voltage: $\pm 1100\,\mathrm{V}$, current resolution: $10\,\mathrm{pA}$
Impedance analyser: Agilent E4991B $f = 1\,\mathrm{MHz}$ bis $3\,\mathrm{GHz}$
Vector network analyser: Anritsu 37397C $f = 45\,\mathrm{MHz}$ bis $110\,\mathrm{GHz}$
On-wafer setup: Cascade Summit 12k   temperature: $-60\,^{\circ}\mathrm{C}$ to $150\,^{\circ}\mathrm{C}$
Bias tee: SHF BT45 $2\,\mathrm{MHz} \leq f \leq 45\,\mathrm{GHz}$, $U_{\max} = 200\,\mathrm{V}$, $I_{\max} = 400\,\mathrm{mA}$
Bias tee: Model 5531 $750\,\mathrm{kHz} \leq f \leq 10\,\mathrm{GHz}$, $U_{\max} = 1.5\,\mathrm{kV}$, $I_{\max} = 20\,\mathrm{mA}$

Symbols and Abbreviations

χ_τ	Steuerbarkeitseffizienz
ε	Often referred to as ε_r
ε'	Imaginary part of $\underline{\varepsilon}$
ε'	Real part of $\underline{\varepsilon}$
ε_0	Dielectric constant
ε_{eff}	Effective relative permittivity
ε_r	Relative permittivity: $\varepsilon' = \varepsilon_0 \cdot \varepsilon_r$
ε_p	Relative permittivity of pore (air)
η_{eff}	Effective FoM
Ω	Maximum volume fraction of inclusions normalized to V
ω	Angular frequency
ω_0	Resonance angular frequency
$\tan\delta$	Dielectric loss
τ	Dielectric tunability
$\underline{\varepsilon}$	Complex permittivity
$\vec{D}$	Displacement flux
$\vec{E}$	Electrical field

$\vec{P}$	Macroscopic polarization
$\vec{p}$	Microscopic polarization
$\vec{r}$	Distance between two coordinates
E	Absolute value of $\vec{E}$
E_C	Electrical coercivity field
E_S	Saturation electrical field
f	Frequency
f_0	Resonance frequency
m_q	Mass of electrical charge
P_R	Reminiscent polarization
P_S	Saturation polarization
Q	Quality factor
q	Electrical charge
q	Volume fraction V_f / V
T_C	Curie temperature
U_B	Bias voltage
V	Volume
V_d	Dielectric volume
V_f	Ferroelectric volume
$W_{s,tot}$	Dissipated energy
$W_{s,tot}$	Stored energy
BST	Barium strontium titanate

BZN	Bismuth zinc niobate
BZT	Barium zirconium titanate
CPW	Coplanar waveguide
CR	Cognitive Radio
EMA	Equivalent medium approximation
FDTD	Finite differences time domain
FoM	Figure of merit
GaN	Gallium nitride
HEMT	High electron mobility transistor
IDC	Interdigital capacitor
IPn	Intermodulation product of the order n
MBO	Magnesium boron oxide
MEMA	Modified equivalent medium approximation
MoM	Method of moments
OFDM	Orthogonal frequency division multiplexing
OIP3	Output intercept point of the order 3
PA	Power amplifier
SDR	Software defined radio
TMN	Tunable matching network

Bibliography

[AKM88]   C. E. Anderson, D. E. Ketchum, and W. P. Mountain. "Thermal conductivity of intumescent chars". In: *Journal of Fire Sciences* (1988), pages 390–410. DOI: 10.1177/073490418800600602.

[Ang+05]   I. Angelov, V. Desmaris, K. Dynefors, P. A. Nilsson, N. Rorsman, and H. Zirath. "On the large-signal modelling of AlGaN/GaN HEMTs and SiC MESFETs". In: *European Gallium Arsenide and Other Semiconductor Application Symposium*. 2005, pages 309–312. URL: http://ieeexplore.ieee.org/abstract/document/1637212/.

[Ast+03]   K. F. Astafiev, V. O. Sherman, A. K. Tagantsev, and N. Setter. "Can the addition of a dielectric improve the figure of merit of a tunable material?" In: *Journal of the European Ceramic Society* (2003), pages 2381–2386. DOI: 10.1016/S0955-2219(03)00139-0.

[BB89]   R. Blumenfeld and D. J. Bergman. "Exact calculation to second order of the effective dielectric constant of a strongly nonlinear inhomogeneous composite". In: *Phys. Rev. B* (1989), pages 1987–1989. DOI: 10.1103/PhysRevB.40.1987.

[Ber78]   D. J. Bergman. "The dielectric constant of a composite material - problem in classical physics". In: *Physics Reports* (1978), pages 377–407. DOI: 10.1016/0370-1573(78)90009-1.

[Bet69]   K. Bethe. "Über das Mikrowellenverhalten nichtlinearer Dielektrika". Dissertation. RWTH Aachen, 1969.

[BMB11]   K. Bathich, A.Z. Markos, and G. Boeck. "Frequency Response Analysis and Bandwidth Extension of the Doherty Amplifier". In: *IEEE Transactions on Microwave Theory and Techniques* (2011), pages 934–944. DOI: 10.1109/TMTT.2010.2098040.

[Bod55]   H. W. Bode. *Network Analysis and Feedback Amplifier Design*. 10th Edition. D. Van Nostrand Company, Inc., 1955.

[Bru35]   D. A. G. Bruggeman. "Berechnung verschiedener physikalischer Konstanten von heterogenen Substanzen. I. Dielektrizitätskonstanten und Leitfähigkeiten der Mischkörper aus isotropen Substanzen". In: *Annalen der Physik* (1935), pages 636–664. DOI: 10.1002/andp.19354160705.

[Bul]   Bulletphysics. Last visit online: 21.10.2016. URL: http://bulletphysics.org.

[BZB10]   O. Blume, D. Zeller, and U. Barth. "Approaches to energy efficient wireless access networks". In: *4th International Symposium on Communications, Control and Signal Processing (ISCCSP)*. 2010, pages 1–5. DOI: 10.1109/ISCCSP.2010.5463328.

[CC16]   Z. H. Chen and Q. X. Chu. "Wideband Fully Tunable Bandpass Filter Based on Flexibly Multi-Mode Tuning". In: *IEEE Microwave and Wireless Components Letters* (2016), pages 789–791. DOI: 10.1109/LMWC.2016.2601280.

[CC41]   K. S. Cole and R. H. Cole. "Dispersion and Absorption in Dielectrics I. Alternating Current Characteristics". In: *The Journal of Chemical Physics* (1941), pages 341–351. DOI: 10.1063/1.1750906.

[CCY05]   D.R. Chase, Lee-Yin Chen, and R.A. York. "Modeling the capacitive nonlinearity in thin-film BST varactors". In: *IEEE Transactions on Microwave Theory and Techniques* (2005), pages 3215–3220. DOI: 10.1109/TMTT.2005.855141.

[Cha+08]   U. C. Chan, C. Elissalde, M. Maglione, C. Estournès, M. Paté, and J. P. Ganne. "Low-losses, highly tunable $Ba_{0.6}Sr_{0.4}TiO_3$/MgO composite". In: *Applied Physics Letters* (2008). DOI: 10.1063/1.2837621.

[CJ79]   R. Cottee and W. Joines. "Synthesis of lumped and distributed networks for impedance matching of complex loads". In: *IEEE Transactions on Circuits and Systems* (1979), pages 316–329. DOI: 10.1109/TCS.1979.1084644.

[CK80]   W. C. Chew and J. A. Kong. "Effects of Fringing Fields on the Capacitance of Circular Microstrip Disk". In: *IEEE Transactions on Microwave Theory and Techniques* (1980), pages 98–104. DOI: 10.1109/TMTT.1980.1130017.

[Cof04]   W. T. Coffey. "Dielectric relaxation: an overview". In: *Journal of Molecular Liquids* (2004), pages 5–25. DOI: 10.1016/j.molliq.2004.02.002.

[Col00]   R. E. Collin. *Foundations for Microwave Engineering*. 2nd Edition. Wiley, 2000, page 37.

[Com]   International Electrotechnical Commission. *Internet of Things: Wireless Sensor Networks*. Last visit online: 21.10.2016. URL: http://www.iec.ch/whitepaper/pdf/iecWP-internetofthings-LR-en.pdf.

[Cou70]   W. E. Courtney. "Analysis and Evaluation of a Method of Measuring the Complex Permittivity and Permeability Microwave Insulators". In: *IEEE Transactions on Microwave Theory and Techniques* (1970), pages 476–485. DOI: 10.1109/TMTT.1970.1127271.

[CS99]   J. H. Conway and N. J. A. Sloane. *Sphere Packings, Lattices and Groups*. 3rd Edition. Springer New York, 1999.

[Dar39]   S. Darlington. "Synthesis of Reactance 4-Poles". In: *Journal of Mathematics and Physics* (1939), pages 275–353. DOI: 10.1002/sapm1939181257.

[Doh36]   W.H. Doherty. "A New High Efficiency Power Amplifier for Modulated Waves". In: *Proceedings of the Institute of Radio Engineers* (1936), pages 1163–1182. DOI: 10.1109/JRPROC.1936.228468.

[DSB08]   G. De Pasquale, A. Soma, and A. Ballestra. "Mechanical fatigue on gold MEMS devices: Experimental results". In: *Symposium on Design, Test, Integration and Packaging of MEMS/MOEMS*. 2008, pages 11–15. DOI: 10.1109/DTIP.2008.4752942.

[Fan48]   R. M. Fano. *Theoretical Limitations on the Broadband Matching of Arbitrary Impedances*. Technical report. Last visit online: 21.10.2016. Massachusetts Institute of Technology, 1948. URL: https://dspace.mit.edu/bitstream/handle/1721.1/5007/RLE-TR-041-08407050.pdf.

[Far+06]   E. A. Fardin, A. S. Holland, K. Ghorbani, E. K. Akdogan, W. K. Simon, A. Safari, and J. Y. Wang. "Polycrystalline $Ba_{0.6}Sr_{0.4}TiO_3$ thin films on r-plane sapphire: Effect of film thickness on strain and dielectric properties". In: *Applied Physics Letters* (2006). DOI: http://dx.doi.org/10.1063/1.2374810.

[Fer+16]   J. Ferretti, S. Preis, W. Heinrich, and O. Bengtsson. "VSWR protection of power amplifiers using BST components". In: *2016 German Microwave Conference (GeMiC)*. 2016, pages 445–448. DOI: 10.1109/GEMIC.2016.7461651.

[Fri14]   Andreas Friederich. "Tintenstrahldruck steuerbarer Mikrowellenkomponenten". Dissertation. Technische Universität Darmstadt, 2014.

[Fri94]   D.A. Frickey. "Conversions between S, Z, Y, H, ABCD, and T parameters which are valid for complex source and load impedances". In: *Microwave Theory and Techniques, IEEE Transactions on* (1994), pages 205–211. DOI: 10.1109/22.275248.

[Gae+09]   A. Gaebler, F. Goelden, A. Manabe, M. Goebel, S. Mueller, and R. Jakoby. "Investigation of high performance transmission line phase shifters based on liquid crystal". In: *European Microwave Conference (EuMC)*. 2009, pages 594–597.

[Gev+96]   S. Gevorgian, T. Martinsson, P. L. J. Linner, and E. L. Kollberg. "CAD models for multilayered substrate interdigital capacitors". In: *IEEE Transactions on Microwave Theory and Techniques* (1996), pages 896–904. DOI: 10.1109/22.506449.

[Gev09]   S. Gevorgian. *Ferroelectric in Microwave Devices, Circuits and Systems*. Springer, 2009.

[Gie+06]   A. Giere, C. Damm, P. Scheele, and R. Jakoby. "LH phase shifter using ferroelectric varactors". In: *Proceedings IEEE Radio and Wireless Symposium*. 2006, pages 403–406. DOI: 10.1109/RWS.2006.1615180.

[Gie+08]   A. Giere, X. Zhou, F. Paul, M. Sazegar, Y. Zheng, H. Maune, J. R. Binder, and R. Jakoby. "Barium Strontium Titanate Thick-Films: Dependency between Dielectric Performance and their Morphology". In: *Frequenz* (2008), pages 47–51.

[Gie09]   A. Giere. "Material- und Bauteiloptimierung steuerbarer Mikrowellenkomponenten mit nichtlinearen Ferroelektrika". Dissertation. Technische Universität Darmstadt, 2009.

[Gon+14]   E. González-Rodríguez, A. Mehmood, Y. Zheng, H. Maune, L. Shen, J. Ning, H. Braun, M. Hovhannisyan, K. Hofmann, and R. Jakoby. "Reconfigurable dualband antenna module with integrated high voltage charge pump and digital analog converter". In: *The 8th European Conference on Antennas and Propagation (EuCAP)*. 2014, pages 2749–2753. DOI: 10.1109/EuCAP.2014.6902394.

[GS01]   J. I. Gersten and F. W. Smith. *The Physics and Chemistry of Materials*. New York: John Wiley & Sons, 2001. ISBN: 978-0-471-05794-9.

[HC60]   B. W. Hakki and P. D. Coleman. "A Dielectric Resonator Method of Measuring Inductive Capacities in the Millimeter Range". In: *IRE Transactions on Microwave Theory and Techniques* (1960), pages 402–410. DOI: 10.1109/TMTT.1960.1124749.

[HS62]   Z. Hashin and S. Shtrikman. "A Variational Approach to the Theory of the Effective Magnetic Permeability of Multiphase Materials". In: *Journal of Applied Physics* (1962), pages 3125–3131. DOI: 10.1063/1.1728579.

[Jeo04]   J. H. Jeon. "Effect of SrTiO$_3$ concentration and sintering temperature on microstructure and dielectric constant of Ba$_{1-x}$Sr$_x$TiO$_3$". In: *Journal of the European Ceramic Society* (2004), pages 1045–1048. DOI: http://dx.doi.org/10.1016/S0955-2219(03)00385-6.

[Kah52]   L.R. Kahn. "Single-Sideband Transmission by Envelope Elimination and Restoration". In: *Proceedings of the IRE* (1952), pages 803–806. DOI: 10.1109/JRPROC.1952.273844.

[Kim+08]   I. Kim, Y. Y. Woo, J. Kim, J. Moon, J. Kim, and B. Kim. "High-Efficiency Hybrid EER Transmitter Using Optimized Power Amplifier". In: *IEEE Transactions on Microwave Theory and Techniques* (2008), pages 2582–2593. DOI: 10.1109/TMTT.2008.2004898.

[Koh16]   Christian Kohler. "Anorganische Barium-Strontium-Titanat-Komposite für die Hochfrequenztechnik". Dissertation. Technische Universität Darmstadt, 2016.

[KSN00]   K. K. Karkkainen, A. H. Sihvola, and K. I. Nikoskinen. "Effective permittivity of mixtures: numerical validation by the FDTD method". In: *IEEE Transactions on Geoscience and Remote Sensing* (2000), pages 1303–1308. DOI: 10.1109/36.843023.

[KUT88]   Y. Kurokawa, K. Utsumi, and H. Takamizawa. "Development and Microstructural Characterization of High-Thermal-Conductivity Aluminum Nitride Ceramics". In: *Journal of the American Ceramic Society* (1988), pages 588–594. DOI: 10.1111/j.1151-2916.1988.tb05924.x.

[Lau+05]   V. Laur, A. Rousseau, G. Tanne, P. Laurent, F. Huret, M. Guilloux-Viry, and B. Della. "Tunable microwave components based on KTa$_x$Nb$_{1-x}$O$_3$ ferroelectric material". In: *European Microwave Conference*. 2005. DOI: 10.1109/EUMC.2005.1608938.

[Lei15]   M. Lei. "The Impact of Composite Effect on Dielectric Constant and Tunability in Ferroelectric–Dielectric System". In: *Journal of the American Ceramic Society* (2015), pages 3250–3258. DOI: 10.1111/jace.13768.

[Len16]   F. Lenze. "Tunable Impedance Matching for High Power GaN HEMT for Optimal Efficiency". Master's thesis. Technische Universität Darmstadt, 2016.

[LL63]   L.D. Landau and E.M. Lifschitz. *Electrodynamics of Continuous Media*. Pergamon Press, 1963, page 43.

[Mau+11]   H. Maune, M. Sazegar, Y. Zheng, X. Zhou, A. Giere, P. Scheele, F. Paul, J. R. Binder, and R. Jakoby. "Nonlinear Ceramics for Tunable Microwave Devices Part II: RF-Characteriziation and Component Design". In: *Microsystem Technologies* (2011), pages 213–224. DOI: 10.1007/s00542-011-1235-9.

[Mau+12]   H. Maune, A. Wiens, O. Bengtsson, W. Heinrich, and R. Jakoby. "Current Status of Tunable RF Components based on Ferroelectric Thick Films for Power Amplifiers". In: *Proceedings of the IEEE Topical Workshop on Power Amplifiers for Wireless Communications*. 2012.

[Mau11]   H. Maune. "Design und Optimierung hochlinearer ferroelektrischer Varaktoren für steuerbare Hochfrequenz-Leistungsverstärker". Dissertation. Technische Universität Darmstadt, 2011.

[Mel+09]   J. Melai, C. Salm, S. Smits, J. Visschers, and J. Schmitz. "The electrical conduction and dielectric strength of SU-8". In: *Journal of Micromechanics and Microengineering* (2009). DOI: 10.1088/0960-1317/19/6/065012.

[Men69]   M. I. Mendelson. "Average Grain Size in Polycrystalline Ceramics". In: *Journal of the American Ceramic Society* (1969), pages 443–446. DOI: 10.1111/j.1151-2916.1969.tb11975.x.

[Mit00]   J. Mitola. "Cognitive Radio An Integrated Agent Architecture for Software Defined Radio". Dissertation. Royal Institute of Technology, 2000.

[Mit95]   J. Mitola. "The software radio architecture". In: *IEEE Communications Magazine* (1995), pages 26–38. DOI: 10.1109/35.393001.

[MSS09]   A. Mueller, J. J. Schneider, and E. Schoemer. "Packing a multidisperse system of hard disks in a circular environment". In: *Phys. Rev. E* (2009). DOI: 10.1103/PhysRevE.79.021102.

[MYJ63]   G. L. Matthaei, Leo Young, and E. M. T. Jones. *Design of Microwave Filters, Impedance-Matching Networks and Coupling Structures.* Stanford Research Institute, 1963, 99ff.

[Nat+05]   J. Nath, D. Ghosh, J. P. Maria, A. I. Kingon, W. Fathelbab, P. D. Franzon, and M. B. Steer. "An electronically tunable microstrip bandpass filter using thin-film Barium-Strontium-Titanate (BST) varactors". In: *IEEE Transactions on Microwave Theory and Techniques* (2005), pages 2707–2712. DOI: 10.1109/TMTT.2005.854196.

[Nik+14]   M. Nikfalazar, C. Kohler, A. Friederich, M. Sazegar, Yuliang Zheng, A. Wiens, J.R. Binder, and R. Jakoby. "Fully printed tunable phase shifter for L/S-band phased array application". In: *IEEE MTT-S International Microwave Symposium (IMS)*. 2014, pages 1–4. DOI: 10.1109/MWSYM.2014.6848295.

[Par+05]   J. Park, J. Lut, S. Stemmert, and R.A. York. "Microwave planar capacitors employing low loss, high-K, and tunable BZN thin films". In: *IEEE MTT-S International Microwave Symposium (IMS)*. 2005. DOI: 10.1109/MWSYM.2005.1516673.

[Par+61]   W. J. Parker, R. J. Jenkins, C. P. Butler, and G. L. Abbott. "Flash Method of Determining Thermal Diffusivity, Heat Capacity, and Thermal Conductivity". In: *Journal of Applied Physics* (1961), pages 1679–1684. DOI: 10.1063/1.1728417.

[Pas89]   F. Paschen. "Ueber die zum Funkenübergang in Luft, Wasserstoff und Kohlensäure bei verschiedenen Drucken erforderliche Potentialdifferenz". In: *Annalen der Physik* (1889), pages 69–96. DOI: 10.1002/andp.18892730505.

[Pau06]   F. Paul. "Dotierte $Ba_{0,6}Sr_{0,4}TiO_3$ -Dickschichten als steuerbare Dielektrika". Dissertation. Albert-Ludwigs-Universität Freiburg im Breisgau, 2006.

[PLL08]   B. Perlman, J. Laskar, and K. Lim. "Fine-tuning commercial and military radio design". In: *IEEE Microwave Magazine* (2008), pages 95–106. DOI: 10.1109/MMM.2008.924969.

[Poz05]   D. M. Pozar. *Microwave Engineering*. 3rd Edition. John Wiley & Sons, Inc., 2005.

[Pre+07]   W. H. Press, S. A. Teukolsky, W. T. Vetterling, and B. P. Flannery. *Numerical Recipes - The Art of Scientific Computing*. 3rd Edition. Cambridge University Press, 2007.

[PZA14]   S. Preis, Z. Zhang, and M. T. Arnous. "Design of a GaN HEMT power amplifier using resistive loaded harmonic tuning". In: *9th European Microwave Integrated Circuit Conference (EuMIC)*. 2014, pages 552–555. DOI: 10.1109/EuMIC.2014.6997916.

[Raa11]   F. H. Raab. "Electronically Tuned UHF Power Amplifier". In: *IEEE MTT-S International Microwave Symposium Digest (IMS)*. 2011. DOI: 10.1109/MWSYM.2011.5972557.

[Rad+14]   A. Radetinac, A. Mani, S. Melnyk, M. Nikfalazar, J. Ziegler, Y. Zheng, R. Jakoby, L. Alff, and P. Komissinskiy. "Highly conducting $SrMoO_3$ thin films for microwave applications". In: *Applied Physics Letters* (2014). DOI: 10.1063/1.4896339.

[Saz13]   M. Sazegar. "Phasedarray-Antennen mit integrierten Phasenschiebern auf ferroelektrischen Dickschichten". Dissertation. Technische Universität Darmstadt, 2013.

[Sch+15]   R. Schmidt, M. Möhring, R.-C. Härting, C. Reichstein, P. Neumaier, and P. Jozinović. "Industry 4.0 - Potentials for Creating Smart Products: Empirical Research Results". In: *Business Information Systems: 18th International Conference, BIS 2015, Poznań, Poland.* Edited by W. Abramowicz. Springer International Publishing, 2015. DOI: 10.1007/978-3-319-19027-3_2.

[Sch07a]   Patrick Scheele. "Steuerbare passive Mikrowellenkomponenten auf Basis hochpermittiver ferroelectrischer Schichten". Dissertation. Technische Universität Darmstadt, 2007.

[Sch07b]   M. Schmidt. "Abstimmbare Anpassnetzwerke auf Basis ferroelektrischer Varaktoren für Mobilfunkanwendungen". Dissertation. Universität Erlangen-Nürnberg, 2007.

[Sch16]   C Schuster. "Hairpin Bandpass Filter With Tunable Center Frequency and Tunable Bandwidth Based on Screen Printed Ferroelectric Varactors". In: *European Microwave Conference (EuMC)* (2016).

[Sco07]   J. F. Scott. "Applications of Modern Ferroelectrics". In: *Science* (2007), pages 954–959. DOI: 10.1126/science.1129564.

[She+06]   V. O. Sherman, A. K. Tagantsev, N. Setter, D. Iddles, and T. Price. "Ferroelectric-dielectric tunable composites". In: *Journal of Applied Physics* (2006). DOI: 10.1063/1.2186004.

[Smy68]   W. Smythe. *Static & Dynamic Electricity,* 3rd Edition. McGraw-Hill Book Company, Inc, 1968.

[STM15]   STMicroelectronics. *Parascan tunable integrated capacitor - STPTIC-27L1.* Rev. 1. Last visited online: 21.10.2016. 2015. URL: www.st.com/resource/en/datasheet/stptic-2711.pdf.

[Tag+03]   A. K. Tagantsev, V. O. Sherman, K. F. Astafiev, J. Venkatesh, and N. Setter. "Ferroelectric Materials for Microwave Tunable Applications". In: *Journal of Electroceramics* (2003), pages 5–66. DOI: 10.1023/B:JECR.0000015661.81386.e6.

[Tan+14]   L. Tang, Y. Bian, J. Zhai, and H. Zhang. "Ferroelectric–Dielectric Composites Model of $Ba_{0.5}Sr_{0.5}TiO_3/Mg_2AO_4$ (A=Ti, Si) for Tunable Application". In: *Journal of the American Ceramic Society* (2014), pages 862–867. DOI: 10.1111/jace.12718.

[Ven07]   O. G. Vendik. "Insertion Loss in Reflection-Type Microwave Phase Shifter Based on Ferroelectric Tunable Capacitor". In: *IEEE Transactions on Microwave Theory and Techniques* (2007), pages 425–429. DOI: 10.1109/TMTT.2006.889348.

[VTZ98]   O. G. Vendik, L. T. Ter-Martirosyan, and S. P. Zubko. "Microwave losses in incipient ferroelectrics as functions of the temperature and the biasing field". In: *Journal of Applied Physics* (1998), pages 993–999. DOI: 10.1063/1.368166.

[Wei+13]   C. Weickhmann, R. Jakoby, E. Constable, and R.A Lewis. "Time-domain spectroscopy of novel nematic liquid crystals in theterahertz range". In: *38th International Conference on Infrared, Millimeter, and Terahertz Waves (IRMMW-THz)*. 2013. DOI: 10.1109/IRMMW-THz.2013.6665423.

[Wei03]   C. Weil. "Passiv steuerbare Mikrowellenphasenschieber auf der Basis nichtlinearer Dielektrika". Dissertation. Technische Universität Darmstadt, 2003.

[Wor]   ON World. *Technology Trends That Are Driving Internet of Things Markets*. Last visited online: 21.10.2016. URL: http://www.onworld.com/news/newsIoT2013.html.

[XMI05]   J. Xu, W. Menesklou, and E. Ivers-Tiffée. "Investigation of BZT Thin Films for Tunable Microwave Applications". In: *Journal of the European Ceramic Society* (2005), pages 2289–2293. DOI: 10.1016/j.jeurceramsoc.2005.03.048.

[YHS93]   K. W. Yu, P. M. Hui, and D. Stroud. "Effective dielectric response of nonlinear composites". In: *Phys. Rev. B* (1993), pages 14150–14156. DOI: 10.1103/PhysRevB.47.14150.

[Zha+10]   Q. Zhang, J. Zhai, L. Kong, and X. Yao. "Percolative properties in ferroelectric-dielectric composite ceramics". In: *Applied Physics Letters* (2010). DOI: 10.1063/1.3514246.

[Zhe11]   Yuliang Zheng. "Reconfigurable Multiband RF Frontends with Tunable Ferroelectric Devices". Dissertation. Technische Universitaet Darmstadt, 2011.

[Zho+10]   X. Zhou, H. Geßwein, M. Sazegar, A. Giere, F. Paul, R. Jakoby, J. R. Binder, and J. Haußelt. "Characterization of metal (Fe, Co, Ni, Cu) and fluorine codoped barium strontium titanate thick-films for microwave applications". In: *Journal of Electroceramics* (2010), pages 345–354. DOI: 10.1007/s10832-009-9580-0.

[Zho12]   Xianghui Zhou. "Prozess- und Dotierungseinflüsse auf $Ba_{0,6}Sr_{0,4}TiO_3$-Dickschichten für steuerbare Mikrowellenkomponenten". Dissertation. Technische Universität Darmstadt, 2012.

[ZZY11]   Q. Zhang, J. Zhai, and X. Yao. "Dielectric and percolative properties of $Ba_{0.5}Sr_{0.5}TiO_3$–$Mg_3B_2O_6$ composite ceramics". In: *Journal of the American Ceramic Society* (2011), pages 1138–1142. DOI: 10.1111/j.1551-2916.2010.04165.x.

Contributions

Publications

First author

[Wie+13a] A. Wiens, M.T. Arnous, H. Maune, M. Sazegar, M. Nikfalazar, C. Kohler, J.R. Binder, G. Boeck, and R. Jakoby. "Load modulation for high power applications based on printed ceramics". In: *IEEE MTT-S International Microwave Symposium Digest (IMS)*. 2013, pages 1–4. DOI: 10.1109/MWSYM.2013.6697548.

[Wie+13b] A. Wiens, O. Bengtsson, H. Maune, M. Sazegar, W. Heinrich, and R. Jakoby. "Thick-Film Barium-Strontium-Titantate varactors for RF power transistors". In: *43rd European Microwave Conference (EuMC)*. 2013, pages 380–383. URL: http://ieeexplore.ieee.org/document/6686916/.

[Wie+14] A. Wiens, O. Bengtsson, C. Kohler, D. Kienemund, M. Nikfalazar, H. Maune, A. Friederich, J.R. Binder, W. Heinrich, and R. Jakoby. "Tunable impedance matching networks on printed ceramics for output matching of RF-power transistors". In: *44th European Microwave Conference (EuMC)*. 2014, pages 496–499. DOI: 10.1109/EuMC.2014.6986479.

[Wie+15] A. Wiens, S. Preis, C. Kohler, D. Kienemund, H. Maune, O. Bengtsson, M. Nikfalazar, J. R. Binder, W. Heinrich, and R. Jakoby. "Tunable In-packageage Impedance Matching for High Power Transistors based on Printed Ceramics". In: *45th European Microwave Conference (EuMC)*. 2015. DOI: 10.1109/EuMC.2015.7345993.

[Wie+16a] A. Wiens, C. Kohler, M. Hansli, M. Schuessler, M. Jost, H. Maune, J. R. Binder, and R. Jakoby. "CAD-assisted Modeling of High Dielectric Contrast Composite Materials". In: *Journal of European Ceramic Society* (2016). DOI: 10.1016/j.jeurceramsoc.2016.10.032.

[Wie+16b]   A. Wiens, C. Schuster, M. Schüßler, R. Jakoby, C. Kohler, and J. R. Binder. "Tunable lumped-element-filter for RF power applications based on printed ferroelectrics". In: *German Microwave Conference (GeMiC)*. 2016, pages 229–232. DOI: 10.1109/GEMIC.2016.7461597.

[Wie+16c]   A. Wiens, S. Preis, C. Schuster, M. Nikfalazar, C. Damm, M. Schuessler, W. Heinrich, O. Bengtsson, and R. Jakoby. "Wideband tunable GaN HEMT module utilizing thin-film BST varactors for efficiency optimization". In: *IEEE MTT-S International Microwave Symposium (IMS)*. 2016, pages 1–3. DOI: 10.1109/MWSYM.2016.7540375.

Coauthor

[Arn+13]   M.T. Arnous, A. Wiens, S. Preis, H. Maune, K. Bathich, M. Nikfalazar, R. Jakoby, and G. Boeck. "Load-modulated GaN power amplifier implementing tunable thick film BST components". In: *European Microwave Conference (EuMC)*. 2013, pages 1387–1390. URL: http://ieeexplore.ieee.org/document/6686925/.

[Arn+14]   M. T. Arnous, A. Wiens, P. Saad, S. Preis, Z. Zhang, R. Jakoby, and G. Boeck. "Evaluation of GaN-HEMT power amplifiers using BST-based components for load modulation". In: *International Journal of Microwave and Wireless Technologies* (2014), pages 253–263. DOI: 10.1017/S1759078714000518.

[Ben+12]   O. Bengtsson, H. Maune, F. Golden, S. Chevtchenko, M. Sazegar, P. Kurpas, A. Wiens, R. Jakoby, and W. Heinrich. "Discrete tunable RF-power GaN-BST transistors". In: *42nd European Microwave Conference (EuMC)*. 2012. URL: http://ieeexplore.ieee.org/abstract/document/6450700/.

[Ben+13]   O. Bengtsson, H. Maune, A. Wiens, S.A. Chevtchenko, R. Jakoby, and W. Heinrich. "RF-power GaN transistors with tunable BST prematching". In: *IEEE MTT-S International Microwave Symposium Digest (IMS)*. 2013. DOI: 10.1109/MWSYM.2013.6697523.

[Fri+14]   A. Friederich, C. Kohler, M. Nikfalazar, A. Wiens, M. Sazegar, R. Jakoby, W. Bauer, and J.R. Binder. "Microstructure and microwave properties of inkjet printed barium strontium titanate thick-films for tunable microwave devices". In: *Journal of the European Ceramic Society* (2014), pages 2925–2932. DOI: 10.1016/j.jeurceramsoc.2014.04.007.

[Fri+15]   A. Friederich, C. Kohler, M. Nikfalazar, A. Wiens, R. Jakoby, W. Bauer, and J. R. Binder. "Inkjet-Printed Metal-Insulator-Metal Capacitors for Tunable Microwave Applications". In: *International Journal of Applied Ceramic Technology* (2015), pages 164–173. DOI: 10.1111/ijac.12362.

[Jos+15]   M. Jost, S. Strunck, A. Heunisch, A. Wiens, A. E. Prasetiadi, C Weickhmann, B Schulz, M. Quibeldey, O. H. Karabey, T. Rabe, R. Follmann, D. Koether, and R. Jakoby. "Continuously tuneable liquid crystal based stripline phase shifter realised in LTCC technology". In: *45th European Microwave Conference (EuMC)*. 2015. DOI: 10.1109/EuMC.2015.7345999.

[Kie+15]   D. Kienemund, C. Nikfalazar M.and Kohlery, A. Friederichy, A. Wiens, H. Maune, M. Mikolajeky, J. R. Bindery, and R. Jakoby. "Temperature dependence of a tunable phase shifter based on inkjet printing technology". In: *German Microwave Conference (GeMiC)*. 2015, pages 150–153. DOI: 10.1109/GEMIC.2015.7107775.

[Kie+16]   D. Kienemund, C. Kohler, T. Fink, M. Abrecht, A. Wiens, H. Maune, J. R. Binder, and R. Jakoby. "A fully-printed, ferroelectric based MIM varactor for high power application". In: *46th European Microwave Conference (EuMC)* (2016).

[Koh+15]   C. Kohler, M. Nikfalazar, A. Friederich, A. Wiens, M. Sazegar, R. Jakoby, and J. R. Binder. "Fully Screen-Printed Tunable Microwave Components Based on Optimized Barium Strontium Titanate Thick Films". In: *International Journal of Applied Ceramic Technology* (2015), pages 96–105. DOI: 10.1111/ijac.12276.

[Lu+14]   D. Lu, Y. Zheng, A. Penirschke, A. Wiens, W. Hu, and R. Jakoby. "Humidity dependent permittivity characterization of polyvinyl-alcohol film and its application in relative humidity RF sensor". In: *44th European Microwave Conference (EuMC)*. 2014, pages 163–166. DOI: 10.1109/EuMC.2014.6986395.

[Man+16]   A. Mani, M. Nikfalazar, F. Muench, A. Radetinac, Y. Zheng, A. Wiens, S. Melnyk, H. Maune, L. Alff, R. Jakoby, and P. Komissinskiy. "Wet-chemical etching of SrMoO$_3$ thin films". In: *Materials Letters* (2016), pages 173–176. DOI: 10.1016/j.matlet.2016.08.038.

[Mau+16a]   H. Maune, M. Jost, A. Wiens, C. Weickhmann, R. Reese, M. Nikfalazar, C. Schuster, T. Franke, W. Hu, M. Nickel, D. Kienemund, A. Prasetiadi, and R. Jakoby. "Tunable Microwave Component Technologies for SatCom-Platforms". In: *Frequenz*. 2016.

[Mau+16b]   H. Maune, M. Nikfalazar, C. Schuster, T. Franke, W. Hu, M. Nickel, D. Kienemund, A. Prasetiadi, C. Weickhmann, M. Jost, A. Wiens, and R. Jakoby. "Tunable microwave component technologies for SatCom-platforms". In: *German Microwave Conference (GeMiC)*. 2016, pages 23–26. DOI: 10.1109/GEMIC.2016.7461546.

[Naq+14]   J. Naqui, C. Damm, A. Wiens, R. Jakoby, Lijuan Su, and F. Martin. "Transmission lines loaded with pairs of magnetically coupled stepped impedance resonators (SIRs): Modeling and application to microwave sensors". In: *IEEE MTT-S International Microwave Symposium (IMS)*. 2014. DOI: 10.1109/MWSYM.2014.6848494.

[Naq+16]   J. Naqui, C. Damm, A. Wiens, R. Jakoby, L. Su, J. Mata-Contreras, and F. Martín. "Transmission Lines Loaded With Pairs of Stepped Impedance Resonators: Modeling and Application to Differential Permittivity Measurements". In: *IEEE Transactions on Microwave Theory and Techniques* (2016), pages 1–14. DOI: 10.1109/TMTT.2016.2610423.

[Nik+13a]   M. Nikfalazar, M. Sazegar, Y. Zheng, A. Wiens, R. Jakoby, A. Friederich, C. Kohler, and J.R. Binder. "Compact tunable phase shifter based on inkjet printed BST thick-films for phased-array application". In: *43th European Microwave Conference (EuMC)*. 2013, pages 432–435. URL: http://ieeexplore.ieee.org/document/6686684/.

[Nik+13b]   M. Nikfalazar, M. Sazegar, A. Friederich, C. Kohler, Y. Zheng, A. Wiens, J.R. Binder, and R. Jakoby. "Inkjet printed BST thick-films for x-band phase shifter and phased array applications". In: *International Workshop on Antenna Technology (iWAT)*. 2013, pages 121–124. DOI: 10.1109/IWAT.2013.6518313.

[Nik+14a]   M. Nikfalazar, Y. Zheng, A. Wiens, R. Jakoby, A. Friederich, C. Kohler, and J.R. Binder. "Fully inkjet-printed tunable S-band phase shifter on BST thick film". In: *44th European Microwave Conference (EuMC)*. 2014, pages 504–507. DOI: 10.1109/EuMC.2014.6986481.

[Nik+14b]   M. Nikfalazar, C. Kohler, A. Friederich, M. Sazegar, Yuliang Zheng, A. Wiens, J.R. Binder, and R. Jakoby. "Fully printed tunable phase shifter for L/S-band phased array application". In: *IEEE MTT-S International Microwave Symposium (IMS)*. 2014. DOI: 10.1109/MWSYM.2014.6848295.

[Nik+14c]   M. Nikfalazar, A. Friederich, C. Kohler, M. Sazegar, Y. Zheng, A. Wiens, J. R. Binder, and R. Jakoby. "Metal-Isolator-Metal Varactor Based on Inkjet-Printed Tunable Ceramics". In: *German Microwave Con-*

176

ference (GeMIC). 2014. URL: http://ieeexplore.ieee.org/document/6775123/.

[Nik+15a]   M. Nikfalazar, C. Kohler, D. Kienemund, A. Wiens, Y. Zheng, M. Sohrabi, J. R. Binder, and R. Jakoby. "Frequency extension of the fully printed phase shifter by paste composite optimization". In: *German Microwave Conference (GeMIC)*. 2015, pages 146–149. DOI: 10.1109/GEMIC.2015.7107774.

[Nik+15b]   M. Nikfalazar, A. Mehmood, M. Sohrabi, M. Mikolajek, A. Friederich, C. Kohler, A. Wiens, Y. Zheng, H. Maune, J. R. Binder, and R. Jakoby. "Low Bias Voltage Tunable Phase Shifter Based on Inkjet-Printed BST MIM Varactors for C/X-Band Phased Arrays". In: *45th European Microwave Conference (EuMC)*. 2015. DOI: 10.1109/EuMIC.2015.7345157.

[Nik+15c]   M. Nikfalazar, C. Kohler, A. Heunisch, A. Wiens, Y. Zheng, B. Schulz, M. Mikolajek, M. Sohrabi, T. Rabe, J. R. Binder, and R. Jakoby. "LTCC Phase Shifters Based on Tunable Ferroelectric Composite Thick Films". In: *Frequenz* (2015). DOI: 10.1515/freq-2015-0082.

[Nik+15d]   M. Nikfalazar, A. Mehmood, M. Sohrabi, M. Mikolajek, A. Wiens, H. Maune, C. Kohler, J. R. Binder, and R. Jakoby. "Steerable Dielectric Resonator Phased-Array Antenna Based on Inkjet-Printed Tunable Phase Shifter With BST Metal-Insulator-Metal Varactors". In: *IEEE Antennas and Wireless Propagation Letters* 15 (2015), pages 877–880. DOI: 10.1109/LAWP.2015.2478959.

[Nik+15e]   M. Nikfalazar, A. Wiens, M. Mikolajek, A. Friederich, C. Kohler, M. Sohrabi, Y. Zheng, D. Kienemund, S. Melnyk, J. R. Binder, and R. Jakoby. "Tunable Phase Shifter Based on Inkjet-Printed Ferroelectric MIM Varactors". In: *Frequenz* (2015), pages 39–46. DOI: 10.1515/freq-2014-0120.

[Nik+16a]   M. Nikfalazar, C. Kohler, A. Wiens, A. Mehmood, M. Sohrabi, H. Maune, J. R. Binder, and R. Jakoby. "Beam Steering Phased Array Antenna With Fully Printed Phase Shifters Based on Low-Temperature Sintered BST-Composite Thick Films". In: *IEEE Microwave and Wireless Components Letters* (2016), pages 70–72. DOI: 10.1109/LMWC.2015.2505633.

[Nik+16b]   M. Nikfalazar, M. Sazegar, A. Mehmood, A. wiens, A. Friederich, H. Maune, J. R. Binder, and R. Jakoby. "Two-Dimensional Beam Steering Phased Array Antenna With Compact Tunable Phase Shifter Based on BST Thick-Films". In: *IEEE Antennas and Wireless Propagation Letters* (2016). DOI: 10.1109/LAWP.2016.2591078.

[Pre+15a]   S. Preis, A. Wiens, D. Kienemund, D. Kendig, H. Maune, R. Jakoby, W. Heinrich, and O. Bengtsson. "Discrete RF-Power MIM BST Thick-Film Varactors". In: *45th European Microwave Conference (EuMC)*. 2015. DOI: 10.1109/EuMC.2015.7345919.

[Pre+15b]   S. Preis, A. Wiens, N. Wolff, R. Jakoby, W. Heinrich, and O. Bengtsson. "Frequency-Agile Packaged GaN-HEMT using MIM Thickfilm BST Varactors". In: *45th European Microwave Conference (EuMC)*. 2015. DOI: 10.1109/EuMC.2015.7346007.

[Pre+16a]   S. Preis, A. Wiens, H. Maune, W. Heinrich, R. Jakoby, and O. Bengtsson. "Reconfigurable package integrated 20 W RF power GaN HEMT with discrete thick-film MIM BST varactors". In: *Electronics Letters* (2016), pages 296–298. DOI: 10.1049/el.2015.4109.

[Pre+16b]   S. Preis, J. Ferretti, N. Wolff, A. Wiens, R. Jakoby, and W. Heinrich O. Bengtsson. "Response Time of VSWR Protection for GaN HEMT based Power Amplifiers". In: *46th European Microwave Conference (EuMC)* (2016).

[Pre+17]   S. Preis, A. Wiens, H. Maune, R. Jakoby, W. Heinrich, and O. Bengtsson. "A highly tunable package integrated 50 Ohm matched GaN transistor module with thin-film BST components". In: *ESA Workshop 2017* (2017).

[Sch+14]   M. Schussler, C. Kohler, A. Wiens, B. Kubina, C. Mandel, A. Friedrich, J. Binder, and R. Jakoby. "Screen printed chipless wireless temperature sensor tag based on Barium Strontium Titanate thick film capacitor". In: *IEEE SENSORS*. 2014, pages 2223–2226. DOI: 10.1109/ICSENS.2014.6985482.

[Sch+16]   C. Schuster, A. Wiens, M. Schüßler, C. Kohler, J. R. Binder, and R. Jakoby. "Hairpin Bandpass Filter With Tunable Center Frequency and Tunable Bandwidth Based on Screen Printed Ferroelectric Varactors". In: *46th European Microwave Conference (EuMC)* (2016), pages 1425–1428.

[Wal+11]   A. Walther, U. Poschinger, F. Ziesel, M. Hettrich, A. Wiens, J. Welzel, and F. Schmidt-Kaler. "Single ion as a shot-noise-limited magnetic-field-gradient probe". In: *Phys. Rev. A* (2011). DOI: 10.1103/PhysRevA.83.062329.

Unpublished Presentations and Workshops

Presenter

[Ben+14]   O. Bengtsson, A. Wiens, S. Preis, R. Jakoby, and W. Heinrich. "Discrete RF-Power Barium-Strontium-Titanate Thick-Film Varactors". Presentation at ESA Noordwijk. Nov. 2014.

[JZW14]   R. Jakoby, Y. Zheng, and A. Wiens. "Funktionskeramiken in der Hochfrequenz-oder Mikrowellentechnik – Grundlagen und Anwendungen". Presentation at *4. Sitzung des Arbeitskreis Funktionskeramik*. 2014.

[Wie15]   A. Wiens. "Impedance matching for high power transistors based on printed ceramics". Presentation at *DPG Tagung*. 2015.

Contributer

[Koh+14]   C. Kohler, A. Friederich, M. Nikfalazar, A. Wiens, R. Jakoby, and J. R. Binder. "Gedruckte Varaktoren für die Mikrowellentechnik". Presentation at *4. Sitzung des Arbeitskreis Funktionskeramik*. 2014.

[Koh+15a]   C. Kohler, B. Kubina, A. Wiens, J. Xu, R. Jakoby, and J. R. Binder. "Fabrication and potential of tunable $Ba_{0.6}Sr_{0.4}TiO_3$-$Mg_3B_2O_6$ composites for microwave applications". Presentation at *Jahrestagung der Deutschen Keramischen Gesellschaft*. 2015.

[Koh+15b]   C. Kohler, A. Wiens, J. Xu, B. Kubina, M. Nikfalazar, R. Jakoby, and J.R. Binder. "Novel fabrication methods for ferroelectric-dielectric composite cerceramics their potential of tailoring the microwave dielectric properties". Presentation at *11th Pacific Rim Conference on Ceramic and Glass Technology*. 2015.

[Mau+12]   H. Maune, A. Wiens, O. Bengtsson, W. Heinrich, and R. Jakoby. "Current Status of Tunable RF Components based on Ferroelectric Thick Films for Power Amplifiers". Presentation at *Proceedings of the IEEE Topical Workshop on Power Amplifiers for Wireless Communications*. 2012.

Supervised Master Thesis

- Christian Schuster - *Tunable Multiband-Filter for Digital Power Amplifiers based on Ferroelectrics*

- Joshua Pauls - *Analog Self-interference Cancellation Network for In-Band Full-Duplex Based on a Finite Impulse Response Filter Structure*

- Ran Hu - *Design and Investigation on Tunable Cross-Coupled Hairpin Filters Based on Ferroelectrics*

- Felix Lenze - *Tunable Impedance Matching for High Power GaN HEMT for Optimal Efficiency*

Curriculum Vitae

Personal Information

Name: Alex Wiens

Date of birth: 20. November 1984

Place of birth: Frunse, Kirgisistan

Education

1996–2005: Konrad-Adenauer-Gymnasium, Meckenheim (Degree: Abitur)

2005–2011: Studies of physics
Johannes Gutenberg-Universität, Mainz (Degree: Diplom Physiker)

since 2012: PhD candidate at the institute of microwave engineering and photonics
Technische Universitaet Darmstadt

Darmstadt, 31.10.2016 ____